Space and Society

Series editor

More information about this series at http://www.springer.com/series/11929

Charles S. Cockell
Editor

Human Governance Beyond Earth

Implications for Freedom

Springer

Editor
Charles S. Cockell
UK Centre for Astrobiology
University of Edinburg
Edinburgh
UK

ISSN 2199-3882 ISSN 2199-3890 (electronic)
Space and Society
ISBN 978-3-319-18062-5 ISBN 978-3-319-18063-2 (eBook)
DOI 10.1007/978-3-319-18063-2

Library of Congress Control Number: 2015939158

Springer Cham Heidelberg New York Dordrecht London

Printed on acid-free paper

Springer International Publishing AG Switzerland is part of Springer Science+Business Media
(www.springer.com)

Preface

Finding a definition for 'liberty' or 'freedom' has never been successful. It is not surprising as these words encompass a vast variety of things. We might speak of freedom from state interference or the freedom to achieve our personal capacities with the help of the state. We might speak of freedom to choose our own government. We can talk of freedom of expression, of thought, of religious belief or freedom of movement. Many of these notions of freedom are interlinked, and some are even inseparable. Many of them can also be examined as separate social or political challenges, confounding the difficulties in finding a coherent self-consistent set of ideas that can be summarised as liberty.

Nevertheless, it is also wise not to allow this complexity to lead one into cultural relativism: that all versions of liberty are just cultural permutations and combinations of these different ingredients, and therefore no human settlement or nation defines a better version of freedom. Freedom of expression, for instance, is not a particular type of liberty legitimately liked by some and not by others. The ability of a human individual to freely articulate their views, in the written or spoken word, on their society or those who govern over them without constant threat of imprisonment or execution is an objectively good type of freedom to expect and demand for all humans. It is right that human individuals should not be denied the opportunity to hear the views of other human beings because those other people have been murdered by a government that happens to find their views disagreeable. If human communities are not to degenerate into collections of contented slaves under the orders of despots, then encouraging independence of mind, and with it freedom of expression, is one objectively desirable form of liberty to pursue at all times and in all places. Freedom of expression is as good on the other side of the Milky Way as it is on Earth.

Maximising liberty in its various manifestations lies at the core of some of the large-scale ideological struggles and even military conflicts of human history. Although it is possible to identify some aspects of liberty that most people can agree are desirable, the conflicting views of what constitutes too much collective oversight and control and what constitutes enough space for individual ideas of the good life can probably never be completely resolved. So long as there are humans with differing views of what constitutes a fully formed notion of liberty then disagreement will exist. And indeed, if all humans did agree on what represents a complete package of human liberty, then ironically it would probably spell the end of free thought and discourse.

The most constructive way to deal with these differences is to build open and free societies where people can debate their ideas of freedom and to construct the political and economic systems that allow for these ideas to be turned into governance and for this governance to change as ideas and opinions alter.

In no place are these challenges more apparent than in outer space. Confronted by lethal conditions, social isolation and the technical complications of supplying the basic needs of air, food and water, extraterrestrial settlements are the locus of a newfound discussion on the nature of liberty. They force us to continue a discourse that began in the emerging democracies of ancient Greece and has continued into the societies of present-day Earth. What sort of freedoms can, or should, people expect on the surface of the Moon when great collective efforts are needed to provide even oxygen to breathe? How can we ensure that people are not driven to desperate depths of depression by the utterly monotonous grey landscape that may sap the sense of self-worth and sense of freedom of mind? Are these questions answered differently on Mars or on an isolated spaceship traveling to a distant planet or even star? These questions merely point at the enormity of the branch of political philosophy that examines extraterrestrial liberty.

On 12 and 13 June 2014 we continued a conversation begun by the UK Centre of Astrobiology and the British Interplanetary Society in 2013. It focussed on the means by which governance structures in space are to be built in a way that maximises the chances for different forms of liberty to flourish. It built on the 2013 discussion which examined the more general idea of what liberty is beyond Earth and what conditions might be necessary for liberty to survive in the extreme conditions of space.[1] Governance in space has been examined before, but there has remained a deficit of discussion about the specific links between governance and human freedom. The chapters presented in this volume are mainly derived from this second discussion at the British Interplanetary Society (Extraterrestrial Liberty II: Human Governance) with added contributions to build a coherent volume.

We, the authors, would like to thank the British Interplanetary Society for supporting the discussion which has led to this collection. We would also like to thank Ramon Khanna and Charlotte Fladt at Springer and Doug Vakoch at the SETI Institute for bringing this book to fruition.

As with our first volume, this book is in one sense a present-day stimulus for discussion. Our motivation was to explore the notion of liberty beyond Earth as a genuinely interesting and important discussion essential to the effort to establish a permanent human presence beyond Earth. However, in another sense, it is also a set of ideas for the benefit of the future inhabitants of the space frontier. Living in these environments, they will have a much better concept of what liberty is and how they

[1]Cockell, C.S. (ed.) (2014). *The Meaning of Liberty Beyond Earth*. Berlin: Springer.

are to realise it than we can probably imagine. Nevertheless, with volumes on liberty, they may at least find themselves with a set of historical ideas that will add to the richness of the totality of available thought. The more ideas there are, the more likely balanced concepts of freedom can be formulated. And, as this volume makes clear, in the lethal extremes of outer space, ideas and concepts of freedom will be in high demand.

Edinburgh, 2015 Charles S. Cockell

Contents

Chapter 1
Introduction: Human Governance and Liberty Beyond Earth

Charles S. Cockell

Abstract Equipped with ideas on the greatest opportunities to maximise liberty in outer space and some of its existential threats, it is possible to proceed to consider how the freedom of space settlers can be incorporated into the governance structures of extraterrestrial settlements. This volume of essays pursues this discussion from an assortment of angles, examining what we can learn from existing and past human communities and political experiences, investigating how free scientific thought and artistic creativity are to be maximised in space and how liberty can be engineered into the very infrastructure of extraterrestrial settlements. Different political and social mechanisms are considered for how impartial laws and governance are to be established. The collection underscores the quantity of information we can use from past experiences of liberty on Earth and the new efforts and ideas that will be needed to prevent the onset of tyrannies in space. Despite the tyranny-prone conditions in the extreme environments of outer space, there are ways in which liberty can be encouraged to thrive and there are approaches that can be formulated long in advance of a permanent human presence beyond Earth.

Keywords Liberty · Governance · Organisation · Constitution · Government

Of the seminal speeches that ring across the ages, there can be little hesitation in saying that Pericles' Funeral Oration, spoken almost two and a half thousand years ago, deserves to rank among the highest. His glorification of the might of Athens and its military prowess may be less appetising in, one would hope, more enlightened times; however, his observations on the virtues of Athens which were the mainspring of the deeds to which he spoke defined a new type of social order:

> Let me say that our system of government does not copy the institutions of our neighbours. It is more the case of our being a model to others than of our imitating anyone else. Our constitution is called a democracy because power is in the hands not of a minority but of the

C.S. Cockell (✉)
School of Physics and Astronomy, James Clerk Maxwell Building, the King's Buildings, UK Centre for Astrobiology, University of Edinburgh, Edinburgh EH9 3JZ, UK
e-mail: c.s.cockell@ed.ac.uk

C.S. Cockell (ed.), *Human Governance Beyond Earth*,
Space and Society, DOI 10.1007/978-3-319-18063-2_1

> whole people. When it is a question of settling private disputes, everyone is equal before the law; when it is a question of putting one person before another in positions of public responsibility, what counts is not membership of a particular class, but the actual ability which the man possesses. No one, so long as he has it in him to be of service to the state, is kept in political obscurity because of poverty. And, just as our political life is free and open, so is our day-to-day life in our relations with each other. We do not get into a state with our next-door neighbour if he enjoys himself in his own way, nor do we give him the kind of black looks which, though they do no real harm, still do hurt people's feelings. We are free and tolerant in our private lives; but in public affairs we keep to the law. This is because it commands our deep respect.

The oration, recorded by Thucydides (1972, p. 145) and delivered in 431 BC after the first year of the Peloponnesian War, was part of Athens' annual tradition to commemorate its war dead. It is remarkable because it admonished its listeners to rally round a set of institutional arrangements and ideas rooted in the notion of freedom. Throughout human history, the masses have been cajoled and galvanised into action by the charisma of dictators or by ideals that are themselves rooted in the power of dictators, monarchs or one-party states. Here we have a speech that justifies sacrifices and implores its listeners to feel pride for the abstract concepts of democracy, a free and open political environment and even letting your neighbour behave how they want without receiving disapprobation, even in a look. Granted, the franchise was not one we would recognise today as very open—women and slaves were excluded from these high-minded ideals—but nevertheless his speech, in its direction, was remarkably modern. Many texts on liberty do not do much more than elaborate on the basic precepts of which Pericles spoke.

The ultimate demise of Athenian power and the fact that two thousand years on we still live in a world where the values that Pericles espoused are enjoyed by a minority of humanity shows how difficult it is to inculcate the abstract ideas of liberty into the minds of people as something worth defending and constructing a society around—how much easier it is to impress them with the physical flesh and bones of a determined dictator.

Yet, this task is indeed difficult, and this observation is made more sobering by realising that in many areas on Earth, there is sufficient availability of water, food and not least air to breathe, thus reducing the number of excuses that despots might have to coerce populations.

How much more difficult would Pericles have found his task if instead of facing the masses in the sun-soaked fields of Athens, he was instead confronted by a multitude gathered under a dome protecting them from the lethal external environment of the Moon? If his oration was an invocation to make good on sacrifices made by people not for military victory, but for the survival of a population faced with instantaneous death caused by depressurisation or the want of liquid water and food, would the abstract ideas of democracy, an open society and allowing your neighbour the life they choose, be so effective?

Some might say that we must answer the question with the response that it must be effective. To deny this would be to accept that all people who venture beyond Earth wander into assured tyranny. To accept that the experiment in democracy and freedom that began in Athens, however many its imperfections, was merely a flourish born in the environment of the Earth and destined to die here would be to consign all hopes for the human settlement of space into the hands of dictators.

In an academic volume such as this one, which considers governance and liberty beyond Earth, there is a certain requirement for objective detachment. First, one must examine the conditions of certain types of liberty beyond Earth and then decide whether they are viable or not. This is a task that a number of us set upon in a previous volume (Cockell 2014).

But there is a more forceful approach as well, namely to recognise the tyranny-prone nature of the extreme extraterrestrial environment and to attempt to find solutions to it. To suggest ways in which we can actively construct governance structures that would allow for a Pericles-like vision of society to emerge and succeed seems an acceptable point of discussion. Academics must be objective, but they should feel no guilt in using their ideas to seek to advance the very liberty that allows them to think freely.

In this volume, we approach the subject of governance and liberty with chapters that examine both the conditions for liberty and the ways in which liberty might be maximised.

Although authoritarianism seems a likely outcome in an environment that is instantaneously lethal, it is not a forgone conclusion. It only seems inevitable because we often extrapolate our experiences on Earth into the extremity of the space environment without modification. However, with prior knowledge of these environments long before we establish human settlements, knowledge gained by robotic craft and humans, we may be in a position to ameliorate the influence of totalitarianism.

Tony Milligan explores the role of social hope and democracy in building extraterrestrial societies and suggests a practical means by which such societies might be constructed. He discusses the use of John Rawls's famous veil of ignorance to consider how we might put in place the constitutional structures to minimise tyranny. That we have time on our side—the opportunity to plan such societies before they are built—gives some hope to his approach. His analysis also uncovers another point that to talk of extraterrestrial liberty now is not premature. It would be easy to ask: Why discuss extraterrestrial liberty when there is no settlement currently in space to worry about this problem? However, the very lack of settlements in space is precisely why we are offered an unusual opportunity in human history to deliberate and discuss human institutional arrangements prior to the settlement of a new environment. This has rarely, if ever, been a possibility before. Even in modern polar stations, governance structures have evolved and developed from haphazard experiences in the early years of polar exploration. We can be fairly sure that whatever solutions are determined for the extraterrestrial case, they will turn out to be in need of modification, but we can take the opportunity to fashion useful ideas and approaches for maximising the chances that the outcome will be felicitous for those inhabiting extraterrestrial environments.

In a chapter that continues the theme of frustrating the extraterrestrial society's slide into tyranny, Charles Cockell discusses what can be done to plan for liberty with a focus on some institutional arrangements but also some very practical every-day considerations. For example, we could approach the task of maximising extraterrestrial liberty by asking ourselves what are some of the most cherished facets of freedom on Earth, or even what facets of freedom are so taken for granted that we generally ignore them. An example could be freedom of movement. On Earth, the ability to walk freely into the outside world without being instantly asphyxiated by the atmosphere (or lack thereof) is a fairly fundamental type of freedom. Indeed, so technically difficult is it to deny people air or the correct atmospheric pressure (without deliberately putting them into a gas chamber) that we rarely address this issue at all in discussions of liberty. However, in extraterrestrial environments, the matter is different and all people must live in pressurised habitats or spacesuits continuously. This inevitably has a curtailing effect on freedom of movement. At once we see a potential conflict. The tyrannically inclined will tend to control the supply and repair of spacesuits and habitats as they will have at their disposal an enormously effective lever for exerting coercive control over a population. However, if we want liberty to be maximised, we should seek to build mass-produced, reliable, easily repaired and easily donned and doffed spacesuits that will enhance the ease with which people can get around, move to other habitats and escape the confines of a settlement. The point is that maximising liberty can be physically engineered as well as incorporated into the more abstract ideas of social and political arrangements.

Of all physical assets that lend themselves to appropriation by tyrants, none is more enticing than land. Land ownership was one of the motivating principles behind John Locke's original ideas for private ownership (Locke 1988) and it has remained a contentious point in thoughts on liberty since then. If there is to be freedom of movement or freedom to establish new settlements independent of other settlements in space, then it is probably essential that land in any given location is not all controlled by a single individual or corporation. Jacob Haqq-Misra discusses ideas for how the acquisition and sale of land can be managed in such a way as to maximise liberty and formulates some new concepts for the private ownership of land.

Despite our best intentions and efforts at engineering, we still have nagging unavoidable challenges of great magnitude in the extraterrestrial environment. The most profound of them may well be access to abundant oxygen. Lack of water and food will certainly, in short order, cause social unrest and potential societal collapse, but a time buffer exists between the point these wants set in and the time to catastrophic famine and thirst. The unavailability of oxygen, however, will kill within minutes and so it commands a fear, a control over the minds of people, that few other commodities can claim. At the same time, it will exert unimaginable attraction to the power-obsessed. Adam Stevens explores the factors that will control the price of this most fundamental of commodities and how access to oxygen will ultimately fashion liberty. He recognises not only the likelihood that the availability of oxygen will influence the culture of liberty, but also that it may be categorically critical to determining whether anyone other than scientists, explorers or others paid to go to the space frontier will want to travel there. Crucial to the settlement of space seems to be creating the means to produce abundant oxygen.

The task that we confront here is not without precedent. Science-fiction writers have made forays, one might even say have been forced, to confront the problems of extraterrestrial liberty. One cannot discuss the unfolding fictitious fortunes of the denizens of the Moon or Mars without discussing their institutional arrangements, both among themselves and in relation to Earth. Inevitably, given the freedom to speculate on any form of these arrangements, science fiction has explored draconian and libertarian extremes and the moderate colonies in between. Stephen Baxter provides us with a comprehensive analysis of the way in which human governance has been explored in science fiction and the lessons that we might learn for liberty. Fictional narratives have one strong advantage: they allow us to explore the possible permutations of human organisation in all their varieties and examine the realities of how they might be avoided or implemented depending on their predicted outcomes. Revolution and liberty have particularly come to the fore when the relationship between Earth and new settlements on our nearest neighbour, the Moon, have been the focus of attention. Many of these narratives provide a stark warning about the influence of tyranny not merely on the unfortunate settlers subjected to it, but also its ultimate influence on the political and economic systems on Earth. Small population sizes, at least in relation to the multi-billion populace of Earth, do not equate to powerlessness when you sit at the top of the gravity well in which those billions live, assuming you control the exit and entry point to that world from the infinite spaces beyond (Dolman 2002). Extraterrestrial liberty is important.

There is much that can be learned from terrestrial societies in solving some of these problems. It is easy to confine past economic and political theories on Earth to the dustbin of history, but there may be approaches and ideas that find fresh impetus in the conditions of the extraterrestrial environment. John Cain explores Marxism in space and examines how his theories can be applied to constructing space settlements. His conclusion is not so much that Marxism will work in space, but that Marx's ideas provide a foundation for an examination of capital and labour relations in space. Even if we decide not to construct a society whose economic relations are operated along Marxian lines, we can at the very minimum attempt to prevent a repeat of some of the more disastrous experiments in Marxism witnessed on Earth.

More modern worlds provide other lessons. From the 19th-century intellectual and 20th-century practical experiments in Marxism, we can move to investigate the late-20th- and 21st-century experiments in virtual worlds. John Carter McKnight explores the lessons from virtual worlds. Built by computer programmers with anarchic and libertarian ideas of what the virtual presence should mean in terms of social organisation, these models were largely rejected by those who occupied these spaces. Instead, he suggests that the lesson we should take from this experience is that it would be worth turning to more communitarian societies, such as indigenous communities, to learn something about how small isolated communities might be governed well while respecting liberty.

Alternative experiences with liberty on Earth provide us with additional examples of how not to do things. Erik Persson explores a particularly pertinent example in the light of the growing presence of private companies in space, namely settlements run by corporations. The history of corporate settlements has not always been a happy one. Slavery and coercion have been common outcomes when corporations have been able to act as both government and operators of profit-driven motives. Persson suggests that these experiences should not make us averse to corporate success and expansion in space, but that a clear separation between those making profits and those making laws needs to be observed. This probably translates into an independently elected civil government whose laws can protect citizens against the worst excesses of corporations. He suggests that the growing role of the private sector in space exploration and settlements makes this an apposite time to discuss extraterrestrial liberty.

For constitutions and other mechanisms of governance to successfully emerge from the legal environment in which settlements are established, there has to be sufficient flexibility. John Rummel explores the existing international legal context in which space exploration occurs, in particular the United Nations Outer Space Treaty, and explores ways in which these structures can be interpreted to successfully balance the need to prevent a free-for-all that destroys extraterrestrial environments and the need to encourage space settlement, particularly by private corporations. His chapter underlines the opportunity to modify existing legal instruments to maximise the effective realisation of liberty and space settlement.

Mukesh Bhatt similarly explores the nature of constitutions and how they will affect the development of settlements in space. In particular, he examines the locus of power and considers how the autonomy of the individual can be maintained. His analysis invokes both present-day constitutions and social arrangements and constitutions that have previously been imagined for extraterrestrial settlements.

Extreme environments on the Earth can tell us a great deal about how policies and competing interests have been managed in the past and how we might avoid these problems in space, or at the very least, use previous governance instruments from these environments to consider how to apply them in space. Like outer space, no person currently lives permanently in the ocean. Thus the management of exploration in this arena might tell us something about space. Lewis Pinault discusses the International Seabed Authority and explains how principles and lessons learned in the implementation of that authority might guide the management of settlements in space with its implications for liberty.

Where do science and art sit within these complex political and economic arrangements? An intriguing and undeniable link exists between liberty and science (Ferris 2010). Scientific culture depends absolutely on a culture that allows freedom of thought, and in the long term this is linked to systems of political institutions that do not coerce and manipulate scientists and the scientific environment. To a great extent, an open society is also dependent on a successful and productive environment of free scientific thought, which is part of the wider health of open, expansive intellectual enquiry. James Schwartz explores how science is to be supported in the extraterrestrial environment to protect this culture of free thinking. He investigates some models of how extraterrestrial authorities might go about funding and generally advancing the culture of scientific work. We might also remember that this task becomes particularly essential in the extraterrestrial environment where the great technical complexity of life-support machines gives authorities the possibility of wielding uncompromising technocratic control over the population if the people do not possess the basic scientific and technical wherewithal to understand the systems that keep them alive. With scientific knowledge comes the ability to question and investigate the decisions of those in power. Scientific knowledge is liberty.

Quite apart from the economic, political and scientific realm, there are ways in which art and culture can be used to mitigate the conformity-driving extremes of the extraterrestrial environment. Annalea Beattie explores how art practices will not only provide a way for the inhabitants of extraterrestrial societies to create new types of art, but they can also be used to positively advance a number of aspects of an extraterrestrial settlement including personal development, social diversity and a sense of inclusion of people in the settlement, the development of freedom of expression and the health of social interactions. Art should not merely be allowed, but its open and free expression should be used as an instrument for good governance and be incorporated into the political instruments of the settlement.

Underpinning the longevity of liberty is general education in addition to its scientific and artistic dimensions. Maximising liberty cannot be done without a well-educated population and mechanisms to pass knowledge to subsequent generations and open the field for discussion. Janet de Vigne investigates the conditions for education in space and how we might incorporate structures of education into the governance of extraterrestrial settlements.

Various common themes are evident within all of these chapters, but two of them stand out.

First, space is not entirely a blank canvas. There is a tendency for people to think that it offers an opportunity to begin afresh, to start a new course for humanity without all its past errors and imperfections. This view is both inaccurate and undesirable. It is inaccurate because many of the facets of the human persona will not change in space. We take our biological and social baggage with us. Society is not a *tabula rasa* to be moulded by the conditions in space into utterly new forms. Different forms indeed, but not categorically separated from their forebears on Earth. It is undesirable because we have over 2000 years' experience of totalitarianism and freedom in governance on Earth, some experiments catastrophically bad, some not so bad. It would be folly to discard these experiences on account of a wild utopian dream and believe that we can begin in ignorance from scratch and build something better. We can learn a vast amount from existing and past human communities that will enable future space settlers to minimise their mistakes and give them the greatest chance of maximising their liberty.

Second, liberty needs work. That isn't a new insight. Individuals on Earth who have cherished freedom of expression have paid considerable prices in social order, money and human lives to advance this single precept of liberty. However, in this volume, we learn a remarkable amount about the specific efforts that are needed to advance liberty on the space frontier. For example, we learn about the intensified problems of separating law-making from profit-making in communities run by private space industries. We learn about the inherent problems of taking libertarianism to extremes in environments where instantaneously lethal conditions demand some collective responsibility. We may want to encourage a deep sense of community precisely for the purposes of saving individual liberty. We learn how art takes on an importance not necessarily appreciated on Earth as an antidote to extreme environmental conditions. We explore how scientific knowledge is necessary in a technically driven extraterrestrial society for people to understand, and even challenge, their authorities. We see how we can consider physically engineering liberty into a society by making technology widespread and easily repaired, thus minimising the potential for control over movement and the production of food, water and air. Among these technologies are spacesuits and oxygen-producing machinery. We can educate people to understand liberty and to think about it. Environmental conditions requiring protean efforts in collective organisation can lead to dictatorship. However, thought, effort and hard work can build liberty-seeking societies even in the most tyranny-prone extremes in space.

The chapters here have a certain feel of prodding in the dark. Each chapter explores its own line of thinking. We are not quite sure how despotism might emerge in space, what aspects of liberty will be the easiest to implement and which the most difficult, and we do not know what aspects of liberty will take centre ground in debates on freedom in locations beyond Earth. These questions can only be resolved by those living on the space frontier. In the meantime, it seems sensible to pursue all paths in the hope that all of them in some way intersect with the real issues that will eventually take shape in the settlements and outposts of the space frontier.

References

Cockell, C. S. (Ed.). (2014). *The meaning of liberty beyond earth.* Berlin: Springer.

Dolman, E. C. (2002). *Astropolitik. Classical geopolitics in the space age.* London: Frank Cass.

Ferris, T. (2010). *The science of liberty: Democracy, reason, and the laws of nature.* New York: Harper.

Locke, J. (1988). *Two treatises of government.* Cambridge: Cambridge University Press.

Thucydides, (1972). *History of the Peloponnesian War.* London: Penguin Books.

Chapter 2
Rawlsian Deliberation About Space Settlement

Tony Milligan

Abstract A genuine political realism about space settlements might allow for the endorsement of a form of settlement democracy, albeit subject to constraints. The value and defensibility of establishing any particular settlement will then depend, in part, upon the constraints which need to be brought to bear in order for survival to be possible. As a more detailed breakdown, Sect. 2.1 will try to show that although there may be strong pressures toward authoritarianism in space, some non-authoritarian political structures may nonetheless provide the most pragmatic candidate option. Section 2.2 will attempt to strengthen this claim by drawing a connection between democracy and social hope, with the latter functioning as a key aspect of any sustainable and worthwhile political culture. Section 2.3 will transition more directly to the context of space and will look at the issue of abortion in a space settlement in order to make a case for constraints upon democratic deliberation. Section 2.4 will argue that Rawlsian deliberation might provide a way for we who are not actually space colonists to realistically theorize such constitutional constraints. Section 2.5 will conclude by suggesting some space-sensitive modifications to the Rawlsian approach.

Keywords Veil of ignorance · Social hope · Rawls · Rorty · Abortion · Democracy

This is a paper about framework rather than constitutional detail, about the possibility of sustaining some form of political realism while deliberating about the unprecedented circumstances of space settlement. I want to suggest that we need a provisional exploration of how to approach the problem of extending political discourse into this new domain if we are to avoid collapsing the discussion into inspired guesswork. The dangers here are perhaps rather obvious. Consider, for example, White Mars (1999) by Brian Aldiss, a text in which constitutional discussions among the political elite of a space settlement skid enthusiastically

T. Milligan (✉)
Department of Philosophy, University of Hertfordshire, De Havilland Campus, Hertfordshire AL10 9EU, UK
e-mail: t.milligan@herts.ac.uk

C.S. Cockell (ed.), *Human Governance Beyond Earth*,
Space and Society, DOI 10.1007/978-3-319-18063-2_2

off-course and result in the creation of a Committee for Evil, dedicated to tackling the latter as a problem of much the same order as sanitation and public health. For those involved, nothing is beyond the scope of sound planning regulations.

To describe this attitude as Aldiss does is already (implicitly) to reject it. But perhaps nothing fits as a model of deliberation when space settlement is at issue. Perhaps everything that we say will be disconnected from political realities. And so, the very idea of deliberation about the political organisation of future space settlements may seem to be not just an outlier of political theory, or an exploration of counterfactuals, but rather a case of the wrong sort of science fiction. The sort that reproduces thinly-disguised versions of present day attitudes and then mistakes them for prophecy. Yet the odd thing here is that there may well be human settlements in at least the nearby regions of space in the not-so-very-distant future, perhaps not cities but something significantly larger than the International Space Station or a polar base, perhaps even stable communities with a reproducing population. Mars, for example, may well be reached before the end of the present century and settled within a timescale of further centuries rather than millennia. Although these lines are written in full knowledge that there have been and will be obstacles, setbacks and tragedies of the sort that might lead any of us to question the value of what is done, the value of exploration, settlement and especially commercial activity anywhere else but here. Such settlements, if or when established, will nonetheless allow for an existence which is very different from life as we know it. And they will also operate with some form of political organization which will be a descendant of our own flawed political practices and institutions. They will, to some extent, bear the stamp of our imbalanced political world just as our institutions and practices bear the imprint of the world of Locke and Rousseau.

And so, from the outset, we are faced with something of a dilemma: evasion of an important fact about the future (which thereby risks sleepwalking into it) and, alternatively, acceptance that fantasy is an acceptable risk (for which the Aldiss scenario stands as a proxy). I want to suggest that this dilemma can, *up to a point*, be overcome although my emphasis here is very much upon the *up to a point.* Beyond a certain level of precision, conjecture of an unmoored sort does begin to take over. Even so, this gives us room in which to operate and the possibility of a *bounded discussion* conducted in a realistic spirit. In what follows, I want to suggest that the realism which can be brought into play need not be identical to authoritarian realpolitik. Rather, and perhaps surprisingly, a genuine political realism about space settlements *might* allow for the endorsement of a form of democracy, albeit subject to various constraints. The value and defensibility of establishing any particular settlement will then depend, in part, upon the constraints which need to be brought to bear in order for survival to be possible.

As a more detailed breakdown, Sect. 2.1 will try to show that although there may be strong pressures toward authoritarianism in space, some non-authoritarian political structures may nonetheless provide the most pragmatic candidate option. Section 2.2 will attempt to strengthen this claim by drawing a connection between democracy and social hope, with the latter functioning as a key aspect of any sustainable and worthwhile political culture. Section 2.3 will transition more directly to the context of space and will look at the issue of abortion in a space settlement in order to make a case for constraints upon democratic deliberation. Section 2.4 will argue that Rawlsian deliberation might provide a way for *we* who are not actually space colonists to realistically theorize such constitutional constraints. Section 2.5 will conclude by suggesting some space-sensitive modifications to the Rawlsian approach.

2.1 The Authoritarian Option

All talk about the politics of space settlements will, of course, be moot if no such settlements are ever likely to be built. If skepticism of the latter sort is correct then what follows *must* qualify as fantasy. And so, in a sense, everything turns upon the assumption that it is overly-pessimistic or simply mistaken about our likely future. Yet deliberation about the politics of space settlements might also be moot for the very different reason that the political structure of such settlements (whatever their size) is bound to be a system of permanent hierarchy and command modelled upon the military, or under the control of 'the company,' or subject to some other authoritarian set-up such as colonial governorship, rather like Hong Kong under the British.

This need not be an 'iron heel' option, complete with armed guards at intersections and tanks upon the lawn. (Hong Kong under British rule did not operate in that way, but it certainly was not a democracy.) Rather, the social norms, policies and practices in place might, *up to a point*, remain liberal but they would not be subject to regular democratic control. This is a familiar option which, in science fiction, is guided to some extent by dramatic necessities, by the need to set up heroes who have enough freedom to be independently-minded, but who also have something authoritarian to rebel against. The option is also, up to a point, driven by a sense of realpolitik in a world where space activities continue to be entangled with military finance as well as state policy, albeit less conspicuously so than in former times and somewhat towards the soft end of military project funding, for example, through the Defense Advanced Projects Agency. The 100 Year Starship project is a case in point. It is backed by DARPA funds. The same is true of some recent research, of a rather more ethically-dubious sort, into suspended animation. (The ethical dubiousness in question stems from the use of animals and from the potential applications of suspended animation for purposes other than interstellar travel.) Should a ready-made structure of authority be needed, or should an ambitiously democratic setup fail under extreme, unearthly conditions, there will always be a known power structure waiting in the wings, ready to take over. Indeed, this is an option which any attempt to colonise another world might have to plan for, with well laid-out criteria for when and how to shift from civilian to military control. It is difficult to imagine the U.S. or China or India failing to devise such protocols as part of any attempt to establish a stable settlement on any scale which truly merited the name.

Cockell (2013) has even suggested that a tendency towards authoritarian control, by either the state or by private financial monopolies, might well be built into the fabric of extraterrestrial societies. Control the oxygen supply in a space settlement and you will have instant control over others, an instant means to enforce their compliance. The extreme vulnerabilities of space, the fact that it can kill you in so many different ways, may generate authoritarian pressures for the sake of sheer survival. There are, however, at least two worrying aspects to such a prospect. The first is the obvious justification problem which it generates. Why, if this is the likely outcome, would anyone want to establish a permanent human settlement on another world?

Here, we might appeal to progress in the future complete with terraforming, planetary engineering and other generations who could enjoy a better life. Or, if skeptical about the latter as a somewhat rosy prospect, we might rely upon the inhabitants of any authoritarian settlement enjoying *enough* freedoms to routinely benefit from the reasonable opportunity of an approximation to a good life. The bottom line here is that this is all that any of us ever have and it might still be possible just so long as social norms remain, up to a point, liberal while politics operates in an authoritarian manner. But here we might wonder about just how broadly liberal any authoritarian system could actually afford to be. After all, the whole point about democracy is not so much the fact that it involves a procedure of majority voting. (We would not regard a political system as functionally democratic if 51 % of the population were allowed to prey upon the other 49 %.) Rather, democracy is not simply procedural but expressive. Democratic practice is, at least in part, a political expression of commitment to liberal norms such as liberty, equality and respect for others. An authoritarian system and liberal social norms would thus be in permanent tension with one another.

And this is where a second worry kicks in. Even if settlement authoritarianism were an acceptable or historically justifiable option, it might not be a stable option. It might be incapable of delivering the security that it promises. On the one hand, authoritarian political organisation might well strengthen a tendency (already conspicuous in Western liberal democracies) towards the sacrificing of liberal social freedoms in the interests of safety and security. On the other hand, even without any such tendency, and precisely because of the tension between undemocratic political systems and liberal social norms, authoritarianism is likely to generate its own counter-culture. (Bakhtin 2009 is the *locus classicus* of the claim that hierarchy of any sort operates in just this manner.)

And while, here on Earth, opposition to authoritarian control can take a long time to feed through into political upheaval, in the intensely vulnerable conditions of space any extensive and deep popular disaffection could prove lethal. Especially so, when mixed with the psychological pressures of prolonged confinement. Terrestrial experiments in enclosed living, even when they have not been disrupted by illicit vodka smuggling, suggest that extreme pressures may ensue and system collapse is never too far away. Prisons are like this too. The Robinson (2009) scenario of an absolute political meltdown with libertarian colonists confronting an uncompromising set of authorities might not be an utterly unrealistic prospect.

Given this, it seems that authoritarianism need not be trumps, even on pragmatic grounds. If it can be made to work, political legitimacy may well be the best policy and that would almost certainly mean some manner of democratic setup or, more precisely, a mixed system with a strong democratic (legitimacy-conferring) component. We do not have more than that on Earth. An element of Authoritarian political control might, therefore, remain part of the mix, but it could not overstep its proper bounds without damaging the overall prospects for social cohesion and settlement survival. Given this, an attempt at deliberation about possible democratic structures in space does not seem to be entirely redundant.

2.2 Social Hope and Unconstrained Democracy

Let us then allow that, with matters approached pragmatically, the optimal form of political organization for a space settlement might not be inherited from some authoritarian practice or institution. Rather, it might involve a variant or component of democracy, adapted in novel ways to unprecedented circumstances. Yet, at this point, we are again in danger of running out of realistic discourse. It might, again, be extremely difficult for us to anticipate such novelty without lapsing into utopian speculation or at least idealization of a sort which is a distance removed from actual political practice. The opening dilemma, the choice between evasion and fantasy, seems to resurface. And here, on the side of fantasy, we have no shortage of Utopias to choose from. We may, for example, like our utopia with a sprinkling of anarchism, like that of Ursula Le Guin's home world, an unyielding society of equally unyielding pioneers. Or we may prefer the more communist-inclined utopia of Tsoilkovsky and of perfectionist Russian philosophy. Matters then become simultaneously a matter of guesswork and the projection of current commitments into the future.

Yet it is important to bear in mind that a rejection of Utopianism, even in the context of space, need not require us to exclude all goals or idealization. We might still work within the confines of the influential vision-preserving distinction drawn by John Rawls between *ideal* and *nonideal theory*. Rawlsian ideal theory describes a 'realistic utopia,' i.e. not a true utopia at all but rather a best possible political arrangement consistent with our human character and interpersonal dynamics and consistent too with the material possibilities of the world we live in (Rawls 1999b, p. 126). Contrastingly, *nonideal theory* deals with more easily attained arrangements but under an important constraint. Such pragmatically-conceived arrangements must not conflict with the pursuit of the ideal. They must keep the possibility of, and perhaps even a move towards, a best-realizable society, in play.

This approach allows for an element of idealization, of a sort which is likely to be integral to any theorizing of the politics of space settlements, yet it is compatible with a form of political realism rather than fantasy. It also combines the advantages of pragmatism with an acknowledgment of the importance of what is sometimes called 'social hope' i.e. hope of the sort which is integral to *any* well-functioning political society. Such hope sustains our sense of the worthwhile. It motivates agents to stand out against the worst of abuses and underpins their willingness to compromise with one another in the belief that additional gains may be made further down the line. As such it is a key part of a political culture. And it is the latter, rather than laws alone, which ultimately sustain democratic institutions.

Yet how we articulate the concept of social hope can vary greatly even if it is done in the context of pragmatic political commitment. Vaclav Havel, one of the key figures of the Eastern European revolutions of 1989, suggested that it was nothing to do with optimism. Rather, hope was all about the belief that, somehow, our actions make sense, that somehow (often in spite of evidence to the contrary) they are justified and/or rational (Havel 1986). For individual dissidents faced to with the task of making tough sacrifices during the long years of Russian dominance, a belief in such justification may have mattered more than faith in a better future to come. Richard Rorty, ever the anti-utopian pragmatist, was a little more forward looking in his articulation of social hope, suggesting that it involves a belief in, and desire for, a future that will be, *in unspecified ways*, better than the past (Rorty 1999). But such formulations may appear rather too modest, too in danger of slipping from a reasonable pragmatism into a standpoint from which the entire exercise of establishing a space settlement may seem too aspirational. They sever the very idea of such hope from any Rawlsian notion of a guiding vision.

This ideal-free hope, although born out of a pragmatic liberal tradition is, of course, familiar from political traditions of another sort. It has been associated (fairly or unfairly) with postmodernism, with the rejection of 'grand narratives,' with laissez-faire neo-liberalism, and with the idea of the open society which is not in pursuit of any special end goal (such as communism, the dictatorship of the proletariat, the truly Christian polity or the realization of Sharia Law) to which individuals might be sacrificed. Along these lines, it may seem best to let matters play out however they will, without either ideal theory or any preconceived political goal for space settlements and to trust instead to the moment-to-moment practical reason of the populus at large. Space may even seem like an opportunity for a more thoroughgoing democracy, an open-ended 'free for all' of popular discussion without any agreed destination and with few or no constraining ground rules for the process of deliberation. Although, formulated thus, an aspiration of this sort looks like it involves a special kind of idealization and perhaps a grand narrative in disguise. (Social hope of a more robust sort may then seem easy enough to send into exile. Keeping it there may be rather harder.)

There is a familiar impulse towards such an idea of unconstrained democracy in neo-liberal thought when the latter is at its most populist, in the idea that we need to cut through the red tape of procedure, and more especially through the constraints imposed by rights legislation, by some overgrown mass of rules and regulations which prevent the sensible will of the majority from being realized at the expense of some or other minority or the disenfranchised poor. (The Tea Party in the U.S., UKIP in the UK and the Abbott government in Australia are recent examples although their attacks upon rights legislation have been taken up by other and more mainstream political forces.) The model of freedom which is on offer in such a rhetoric of liberty is decidedly negative freedom. It is *freedom from* rather than *entitlement to* some form of enabling which may place a call upon community resources or the taxpayers. Indeed, it is freedom with distributive justice out of the picture. Neo-liberal equality of this sort is largely political rather than economic, in spite of the fact that wealth grants political privileges which poverty does not. At more somber moments, the periodic resurgence of such populism may lead us towards the suspicion that one of the oldest objections to democracy (the Platonic objection) may have been correct all along: it tends towards the rule of the uninformed and the basely-manipulated, it drags us back to where the shadow-play of illusions is at its strongest.

But while there is usually an element of political contrivance in the deployment of such appeals, we might nonetheless allow that, in theory at least, a better-motivated and no-holds-barred democratic free-for-all might proceed in an informed manner, without manipulation of sorts which are familiar within our terrestrial party systems. This is an option which also has the advantage of building public deliberation strongly into the picture (one of the dominant moves within contemporary democratic theory). The idea of an unconstrained democratic free-for-all is also realistic at least in the sense that it is an option which is likely to tempt any group of colonists who are left to their own devices or placed beyond the geographical reach of terrestrial authority. Whatever agenda or instructions they are sent with may seem distant, another world may be there for the making with no political maps on tables and no need to comply with what has been determined elsewhere. Here, we may think of the rallying cry of Kim Stanley Robinson's fictional colonists in once they are safely beyond control: 'I don't think we should pay any attention to plans made for us back on Earth!' (Robinson 2009, p. 77). The lure of this thought is considerable.

But one of several difficulties for any such unconstrained option is that it simply does not yield a model of stable democratic government, and without the latter authoritarianism may result. And the reason why it is a poor model for stability is nothing to do with public engagement or the spread of deliberation beyond the bounds of political elites. (Generally, these are good-making features of a democratic system, features upon which legitimacy may depend. Referenda, voter initiatives and public for a contribute to a healthy political culture.) Rather, the problem is that a democratic free-for-all is unlikely to yield a reasonable political consensus. As soon as colonists come up against a divisive issue, any prospect for such a consensus is likely to collapse. This is not a new point, albeit it is applicable in the new setting of space. It is the central insight of classical republican political theory. When, for example, Rousseau appealed to the virtues of the people and to the *popular will* he was not appealing to an unconstructed *given*. Rather, his appeal was to a built-agreement which was shaped by a suitable set of civic institutions, practices and ground rules of a sort which would encourage social solidarity and promote belief in the reality of a *common good*, a shared condition of flourishing for which hopes might then be entertained.

2.3 Constrained Deliberation

As a case in point, an illustration of how easily a socially-unconstrained political discussion may lead to intractable division, we might consider the issue of abortion. This is, rather notoriously, a point of disagreement which pits established ideologies against one another. As such, finding a way to deal with this issue, without making divisions worse, may be a litmus test for the viability of an approach towards deliberation about the politics of space settlement.

To recap on the familiar arguments: pro-choice positions typically proceed by appeal to the accepted entitlement that we have, under liberal norms of respect, to bodily integrity. Crudely put, we own our bodies. But ownership here is too weak a concept because it is too easily countermanded. After all, your house may be subject to compulsory purchase but nothing similar applies to your body. Nobody can use or harvest your organs without consent, even if doing so would not kill you but would save several other lives. Nor are the courts entitled to sentence rapists to be raped or to mandate the removal of limbs as punishment, even in the case of the most serious and deplorable of offences. Bodily integrity is, in any familiar liberal context, non-negotiable. It is basic to our conception of respect for the individual, indeed it is integral to our terrestrial understanding of what it is *to be* an individual. Under ordinary circumstances it also extends beyond life. We continue to bury useful organs and allow people to die rather than enforcing their seizure.

Anti-abortion arguments typically bypass such talk about bodily integrity or else they suggest that it is trumped by the sacredness of (innocent) human life or, in less religious terms, by a fundamental right to life which kicks in at different points in time depending upon the version of the pro-life position in question: conception, the point of viability or with the development of a central nervous system. The discussion is then shifted onto the grounds of developments in medical technology. This setting aside of the issue of bodily integrity does, at least in some cases, generate anomalies for opponents of abortion who are not motivated by religious arguments because in other contexts they too abide by the norms in question and so abortion is regarded as an exception, a special case which is special because the stakes are so high.

What consideration of these matters in the context of space settlement does is to overlay an additional set of equally conflicting considerations on top of the already established lines of division. Against abortion, it may be pointed out that becoming pregnant and carrying to term may be extremely difficult under altered gravitational conditions (Ronca 2007; Philips 2012). Abortion might then militate against group survival in ways which are more than symbolic. However, in support of a right to choose, the idea of an *in principle* opposition to abortion at whatever stage may seem unrealistic given the fact that there will always be constraints upon the carrying capacity of space settlement infrastructure. Ideology will not alter the number of people who can actually be kept alive, under tolerable conditions, at any given time. We might then be inclined to agree upon two key points. Firstly, that the basic problem here is the carrying over of too much ethical baggage from Earth. And this is quite different from the romantic claim that ethics can start from scratch, that tables of values can be written in the morning and erased in the evening, or that terrestrial ethical norms will simply become an irrelevance elsewhere. Rather, it is the more restricted point that one or both sides in such debates may need to shed at least some of their more *situation-sensitive* conceptions of what counts as right and wrong if enough of a consensus about workable compromise options is to be secured.

Secondly, and more generally, getting political agents to surrender, depart from or otherwise qualify some of their terrestrial attitudes may be all the more important if, as has been suggested elsewhere, deliberation about ethics and/or politics in space is *generally* or regularly liable to be dilemmatic (Milligan 2011, 2015; Schwartz 2014a, b). And so the abortion issue may turn out to be typical rather than atypical of the more difficult social issues that space settlement is likely to generate. In the context of space politics, dilemmatic may equate with *dangerously dilemmatic*. And here, the suggestion is not that diversity is bad, in space or anywhere else, but rather that ways of avoiding utterly intractable disputes about *fundamental* policy (such as settlement reproduction) may be a basic requirement for survival.

Democratic political deliberation may then need to proceed under the influence of mechanisms, formal and otherwise which can help to promote a workable level of compromise and consensus. More precisely, there may need to be constitutional constraints upon practical deliberation in order to help us screen-out some of the familiar terrestrial preconceptions which would prevent a workable compromise from emerging. *In extremis*, arguments from one section of the political community that another section are vermin should not make it onto the policy agenda. And this is rather different from merely suggesting that such arguments should merely be rejected whenever raised. This, in a sense, is continuous with some already existing aspects of terrestrial political practice which screen out socially corrosive arguments. In other words, democratic political deliberation as we know it is *never* the free-for-all towards which a neo-liberal populism aspires. There are always constraining rules and norms built into discussions, rules and norms which are set formally by acceptable practice and informally by the prevailing political culture. Such constraints can, of course, be a mechanism for discursive conservatism but they can also be an important mechanism for social solidarity.

2.4 The Rawlsian Device for Constitution Making

One way to think about such mechanisms (not the only way, but a useful way) is in terms of a settlement constitution which places restrictions upon what community members can enforce and upon which individual entitlements can be modified by regular (non-constitutional) law. However, if we are to engage in realistic deliberation about what might be included within such a constitution then we too might need to abide by some screening mechanisms, albeit not necessarily the same ones which would shape political practice within a space settlement, and not necessarily those which might be suitable for terrestrial politics as it is currently practiced.

The constraining option that I have in mind is a version of the Rawlsian 'veil of ignorance' (Rawls 1999a, b). John Rawls invites us to imagine an original position, a situation where agents engage in deliberation prior to the establishment of a political community, and where, as self-interested agents, they would discuss possible social arrangements and basic principles of justice without knowing where they would be situated in the society that results. All these agents would take into this original condition would be their self-interest, their bare rationality and rudimentary sorts of social knowledge of a sort which is basic to an understanding of the bounds of the possible. All special attachments and affiliations, even knowledge about the agent's own gender, ethnic background and physical characteristics, are to be bracketed out of the picture.

Here, in addition to what is made explicit by Rawls, I want to suggest that there is an additional implicit commitment to the deliberators being in possession of basic knowledge about the material environment in which they live. (For example, whether or not the local gravity, levels of radioactivity or chemical atmosphere are such that they would pose problems for reproduction.) The rationale for this extension of Rawls is much the same as the rationale for including the basic social knowledge clause. Rudimentary environmental knowledge is necessary because the veil of ignorance is a device for building an ideal theory and not a true utopia. The results of deliberation must therefore abide by the principle that *ought* implies *can* and we will simply not know what can be done unless we also know various things about the material circumstances in which we are situated. While explicit provision for this sort of knowledge is usually absent from accounts of the Rawlsian original position, it is only missing because of the assumption that the world in question will be our own.

Behind such a Rawlsian veil of ignorance, in order to secure our own well-being, agents would agree to the fullest *possible* system of liberties which was consistent with liberty for all. Or so Rawls believes. Moreover, he holds that agents would also try to minimize the harms to which they might be exposed by agreeing to inequalities in the distribution of goods only if they happened to improve the position of the least well-off. (Always bearing in mind the possibility that the least well-off might be themselves.) This is the so-called 'difference principle' and it will figure again below. Backing up claims about political principle will require us to show that agents under such veil of ignorance circumstances might well agree to the principles in question. One strong advantage of this approach, at least as far as the generation of compromise and consensus goes, is that it provides a way of making sense of what individual rights we should acknowledge without appeal to classical Lockean political theory and to the idea of natural rights, the idea that rights are in some way built into the fabric of nature. Rather, although there can still be ethical truths, what makes them true may now be understood in broadly *constructivist* rather than *naturalist* terms, as a matter of what suitable, unbiased rational agents would ordinarily agree to and what they would ordinarily accept as a reason for action. This may not yield a conception of ethics which plumbs the metaphysical depths but, here, we may be inclined to say 'Who could ask for anything more?'

Critics have, however, pointed out that, in spite of its goal of circumventing bias, the Rawlsian veil of ignorance works suspiciously like a device for turning out liberal principles. And, in a sense, that is exactly how it does operate. Rawls does not claim comprehensive value neutrality but instead accepts that the principles of justice which his thought experiment yields will always depend upon how the original position is actually set up. It will depend, for example, upon exactly what sort of basic or minimal knowledge and commitments are taken behind the veil of ignorance and what is pared away. Change the starting conditions and a different result may then ensue. However, for Rawls, the choice of initial conditions is not arbitrary. Instead, the favored conception of the original position is to be shaped by a method of 'reflective equilibrium.' The method in question appeals to intuitions in the full knowledge that they are simultaneously unreliable *and* indispensable. We need them, because deliberation must begin somewhere, but we also know that they can be misleading or unfair. (Intuitions about race and justice in the antebellum-South might not have been an ideal guide to action.) And so, instead of trusting our intuitions blindly, we need to use them in order to shape our theories and then use our theories, and requirements of consistency, in order to shape our understanding of which intuitions are reliable or at least plausible. That is how ethical and political theory works.

What makes this entirely framework problematic when reasoning about terrestrial politics is that our intuitions and theories will always be skewed by the actual knowledge that we have about who we are and about where we (and those we most identify with) are situated in society. It will yield a semblance of objectivity where there is none. However, with regard to any future space settlement this would not so obviously be the case. Indeed, none of us knows anything about our situation in such a future society because we do not happen to have one. We are deliberating about the future of others, perhaps not very distant others but perhaps they may be distant enough to counteract certain kinds of worrying partiality.

What we might then expect from agents behind a veil of ignorance, who are not utterly wedded in advance to any doctrinaire position but who are committed to establishing a *sustainable* space community, is a reasonable acceptance that viable principles of justice and ethics require a sustaining material context. However, the material context for certain kinds of familiar ethical standpoints might simply not be available. Determining whether or not it was would be a quasi-empirical matter, dependent upon basic environmental knowledge. Such standpoints, even if they form part and parcel of our everyday terrestrial outlook, should not then qualify as part of an ideal theory because movement towards them in the given otherworldly situation might be well-motivated but socially destructive. As a matter of clarification, this is *not* a reversion to what Stephen Baxter calls the 'cold equations' of utilitarianism, but rather a basic adequacy condition for political realism which applies as much on the Earth as it does elsewhere (Baxter 2014). Even here, the only viable ethical options are ones which we can actually live by. And, as below, so above, or at least 'so elsewhere.'

This will lead to concessionary moves from differently motivated individuals. For example, while I happen to be a fairly radical egalitarian with regard to terrestrial politics, I should perhaps be ready to concede that a reasonable, non-doctrinaire, agent deliberating behind a veil of ignorance might well hold that such egalitarianism would not be viable as part of the ideal theory *for some particular kind of world* in the light of the information that we know about it. And this is the same as saying that we would need to know what any given world was actually like before we could say *for certain* that egalitarianism was possible there. Similarly, I hold very firm views about abortion here on Earth, along the lines of the bodily integrity argument and, with Rawls, I suspect that agents deliberating behind a veil of ignorance about *terrestrial* circumstances would hold to some variant of the same position. But it is by no means obvious that they would continue to do so irrespective of all possible alterations that might be made to the basic social and environmental knowledge which they had to work with. And to say this is to accept (reluctantly) that a liberal attitude towards bodily integrity might not be viable under *all* of the varying conceivable circumstances which a space settlement might have to face. Whether or not any particular rights of bodily integrity could reasonably be sustained as part of the situation specific ideal theory of justice would depend again upon the basic *situational* information that deliberators have to go on and the intractability of the problems faced. It is not something which could be determined a priori, in advance and independently of both social and environmental knowledge.

If correct, this will have serious repercussions for what we, in the here and now, aspire towards. The very fact that abortion rights might then have to be compromised, and perhaps indefinitely so, if a settlement is to survive (particularly in the case of intractably remote or isolated settlements) could shape our attitude towards how worthwhile the creation of a space settlement *in such places* might be and whether or not we terrestrial agents should ever support it and thereby condemn future humans to a partly-illiberal predicament. The creation of any settlement with a built-in requirement for norms which are radically different from our own would be a serious matter and might be difficult to justify. Conversely, those who are thoroughly committed to an unbending anti-abortion stance on Earth might have to concede its impracticality under at least some conditions of *sustainable* space settlement. Failure to make such a concession would surely involve a divergence from what could be affirmed by any non-doctrinaire agent deliberating behind a veil of ignorance but given unfavorable basic knowledge about social and environmental realities, reproductive options and population limits. And this too might give the advocates of a committed anti-abortion stance a reason to consider the advisability of establishing space settlements in a least some places, under some circumstances, because of the tragic dilemmas that their creation might help to generate.

2.5 Qualifying the Veil of Ignorance

In a sense, I am suggesting that in response to the dilemma which involves having to choose between evasion about the future or the risk of fantasy, we might run a Rawlsian veil of ignorance thought experiment, but only in a modified form and as an orphaned device, one which can survive the demise of various background commitments which helped to bring it into being. This is all the more important because, were we to be consistent Rawlsians through and through, it is not immediately obvious that we would be in a position to engage in space settlement at all or in a position to apply the Rawlsian machinery in order to make sense of the relevant principles of justice. Other attempts to appeal to Rawls have been equally tentative and guarded. Schwartz (2014b) has recently pointed out that human activity in space might tend to increase inequalities without improving the position of the least well-off. In which case, they would conflict with the difference principle. And arguably, this has, at least sometimes, been the case although Schwartz holds that there is an available and sufficiently Rawlsian solution.

Perhaps more of a problem for a consistent Rawlsian is the fact that Rawls always had strong reservations about using the veil of ignorance in order to reason about justice in relation to other creatures and future generations because they could not participate in here-and-now deliberation (or any deliberation at all). And so their inclusion must always compromise the veil of ignorance by idealizing it to a greater extent than Rawls was prepared to countenance. Such an approach also generates a problem of authority. If the reasoning agents are not pared-back and more rational versions of ourselves then why should *their* agreements be authoritative for *us*? Even so, the reasons for such exclusions are notoriously controversial and Rawls may well have been wrong about exactly what his own theory implied in special cases.

Finally, I will suggest that there is a significant adjustment which needs to be made to the veil of ignorance in light of the kinds of intractable vulnerability in space to which Cockell (2013) has drawn attention. And here, I do mean an adjustment and not a simply an extension of the sort which is required to secure the inclusion of basic environmental knowledge. The adjustment concerns the community-mindedness of deliberators as opposed to their motivation by self-interest. While the right kind of ethic for space settlement might be, as Arnould (2011) has suggested, a 'frontier ethic,' it would be misleading to press familiar homesteading analogies with familiar representations of the U.S. frontier too far. And not simply because of the familiar ethical criticisms of colonizing already-inhabited territory at the expense of an indigenous population. While the U.S. frontier encouraged an ideology of sturdy individualism, albeit at odds with the overall realities of community dependence, the notion of the radically independent agent, the lonesome cowboy, makes very little sense in the space context. Within a settlement, at least during any early phase of colonization (i.e. prior to mass communities, multiple cities or extensively inhabited worlds) there can be no space cowboys unless they happen to be maintained at the community's expense, as a luxury. And this is a rather unlikely scenario. The range of human interdependencies, from the skills that agents would need to acquire through to the roles they would be need to play for the sake of group survival, is far too complex and tightly interwoven to allow room for drifters and agents who wish merely to 'do their own thing' underneath the starry skies above.

When we reflect upon considerations of this sort, we can begin to see just how strong the pressures towards authoritarianism might be. We may also be well placed to recognize that the Rawlsian decision to take only the motivation of self-interest behind the veil of ignorance, rather than a mixture of self and other interest, may not yield enough of an understanding of the kinds of solidarity and community identification which are likely to be necessary for group survival. The difficulty of imagining such a group-oriented mindset, a real difficulty from the standpoint of a liberal and strongly individualist culture, can be addressed, *up to a point*, if we assume that the deliberating agents who operate behind the veil of ignorance are allowed to see themselves from the outset, not just as contracting individuals but as the bearers of a shared humanity. Although what this involves may be a subject of dispute, the option which I have favored elsewhere is one which closely connects an understanding of our humanity with a sense of our shared vulnerability (Milligan 2014). And this leads me to suggest, tentatively, that the right kind of initial ethic for the space frontier may well be, as Jacques Arnould claims, some qualified form of 'space humanism.' (Arnould 2011, pp. 121–131). Or, at the very least, it may draw something from the latter. And the relevant conception of our shared humanity may, in turn, involve an orientation towards shared origins, and thus a sense of connection to a distant Earth. It may even involve a sense of connection to earlier generations of deliberators such as ourselves, reasoning about space, but doing so in the dark and under all manner of epistemic constraints.

Bibliography

Aldiss, B., & Penrose, R. (1999). *White mars*. London: Little Brown.

Arnould, J. (2011). *Icarus' second chance: The basis and perspectives of space ethics*. New York: Springer.

Baxter, S. (2014). The cold equations: Extraterrestrial liberty in science fiction. In C.Cockell (Ed.), *The Meaning of Liberty Beyond Earth* (pp. 13–31). Heidelberg, London, New York: Springer.

Bakhtin, M. (2009) *Rabelais and his world*. Bloomington: Indiana University Press.

Cockell, C. S. (2013). *Extra-terrestrial liberty*. Edinburgh: Shoving Leopard.

Havel, V. (1986). 'The politics of hope' in Havel. *Disturbing the Peace*. 163–205.

Milligan, T. (2011). Property rights and the duty to extend human life. *Space Policy, 27*(4), 190–193.

Milligan, T. (2014). Fear of freedom: The legacy of Arendt and Ballard's space skepticism. In C. Cockell (Ed.), *The Meaning of Liberty Beyond Earth* (pp. 33-45). Heidelberg, London and New York: Springer.

Milligan, T. (2015). *Nobody owns the moon: The ethics of space exploitation*. North Carolina: McFarland.

Philips, R. W. (2012). *Grappling with gravity: How life will adapt to living in space*. New York: Springer.

Rawls, J. (1999a). *A theory of justice* (2nd ed.). Oxford: Oxford University Press.

Rawls, J. (1999b). *The law of peoples*. Cambridge: Harvard University Press.

Robinson, K.-S. (2009). *Red mars*. London: Harper Voyager.

Ronca, A. E. (2007). Effects of spaceflight and altered gravity on reproductive processes of female mammals. *Gravitational and Space Biology Bulletin, 20*(2): 97–98.

Richard, R. (1999). *Philosophy and social hope*. London: Penguin.

Schwartz, J. S. (2014a). Prioritizing scientific exploration: A comparison of the ethical justifications for space development and for space science. *Space Policy 30*(4), 202–208.

Schwartz, J. S. (2014b). Fairness as a moral grounding for space policy. C.Cockell (Ed.), *The Meaning of Liberty Beyond Earth* (pp. 69–89). Heidelberg, London and New York: Springer.

Chapter 3
Extraterrestrial Liberty: Can It Be Planned?

Charles S. Cockell

Abstract The very phrase 'planning liberty' seems a contradiction in terms. Yet, there are ways, through physical architecture, legal systems and social norms, that an environment can be created that maximises the potential for individuals to both have a sphere of activity in which they can enjoy non-interference and in which they can maximise their potential to realise their goals. In space, the environmentally extreme conditions warrant a search for all means to maximise liberty in its various manifestations against the extreme tyranny-prone nature of the environment. An example of a liberty-driven technological innovation is a reliable, plentiful and easily donned spacesuit that maximises the potential for freedom of movement in an otherwise lethal environment. Another is the development of fast, reliable spacecraft for moving goods around the solar system that minimise the effects of economic isolation and maximise the potential for efficient free trade. An example of liberty-driven administrative planning is free access to all information about oxygen supply and demand to prevent tyrannical control of oxygen supplies. When constituted deliberately around the desire to maximise liberty, there are technological, economic, political and cultural ways in which tyranny in outer space can be held at bay. I suggest a 'liberty calculus' as one way to approach this challenge.

Keywords Liberty · Planning · Freedom of movement · Extremes · Economics

3.1 Introduction

The title of this essay is inherently suspect because as soon as we discuss planning liberty, we are left with a series of troubling questions: Whose version of liberty are we planning? Who gets to check the planning is being done correctly? Who gets to decide what plans are no longer servicing liberty and should be dismantled?

C.S. Cockell (✉)
UK Centre for Astrobiology, School of Physics and Astronomy, James Clerk Maxwell Building, The King's Buildings, University of Edinburgh, Edinburgh EH9 3JZ, UK
e-mail: c.s.cockell@ed.ac.uk

C.S. Cockell (ed.), *Human Governance Beyond Earth*,
Space and Society, DOI 10.1007/978-3-319-18063-2_3

On a more principled level, therefore, I will say from the outset that liberty is something to be constantly and openly discussed.[1] Relegating it to a planning committee is likely to kill it before it has a chance to flourish, and nowhere is this more probable than in outer space. Surrounded by an environment which lacks naturally available food, liquid water and oxygen, to divest the development of a settlement to a group of people who have been tasked with 'planning liberty' will surely lead to disaster, no matter how well intentioned the committee may be at the outset. What then do I mean by the title of this essay?

There are aspects of the space environment that are predictably tyrannous and they are despotic in a way which evades specific definitions of liberty. The extreme nature of space and the fact it forces all individuals into habitats has within it the potential for the central control of those people and the supplies of their basic requirements for sustenance on levels unseen on Earth.[2] The inability to open a door and move freely outside or just to escape for a few hours on one's own provides the opportunity for people to control freedom of movement rarely afforded to the most coercively inclined terrestrial dictator. The isolation of settlements on distant planets, moons, asteroids or spaceships generates extreme economic separation. This remoteness, forced upon settlements not by determined oppressors, but by the laws of physics, nevertheless plays into the hands of the autocrat by reducing opportunities for trade, the movement of workers and the free exchange of capital. With these restrictions emerges the possibility of a population becoming utterly submissive to the authorities that govern the settlement.

These aspects of tyranny, and there are others besides, are side effects of the extreme isolation and environmental conditions in which settlements become established. There is much to argue about concerning the detail of liberty in space which might lead us to the general view that we should not attempt to plan liberty,

---

[1]I am assuming here that the advance of science and the movement of people beyond Earth will not result in some entirely new type of rational human being who lies beyond the conflicts of liberty we see on Earth today. There is a tendency, particularly in Enlightenment literature, to make assertions that such utopias exist (at least on Earth). A classic view by de Condorcet (1979, p. 179), for example: 'The time will therefore come when the sun will shine only on free men who know no other master but their reason; when tyrants and slaves, priests and their stupid or hypocritical instruments will exist only in works of history and on the stage.' It is likely to be safer to assume that tyranny will always exist and then attempt to mitigate it than to hope it will go away completely. Furthermore, if we accept that there will never be an objective agreement on what liberty is (and maybe such a point would in fact spell the end to free thought), then we must find ways to allow a healthy disagreement and discussion on what constitutes tyranny at any given time in human history.

[2]The excuse for control derives from facing a type of 'enemy' and in some ways the psychology that results is reminiscent of the collectivism that war causes, as Millar (2006) recognised long ago (p. 180): 'There is no situation in which a body of men are so apt to run into disorder, as in war; where it is impossible that they should cooperate, and preserve the least regularity, unless they are united under a single person, empowered to direct their movements, and to superintend and control their several operations.' If the common enemy is the extraterrestrial environment, and existence is adeptly turned into a collective war against this environment by the managers of a settlement, then central control follows.

lest the planning falls into the hands of tyrants. However, if we agree with the assertion that 'outer space is tyranny-prone and we want to prevent populations falling prey to despots', then there are aspects of liberty that we might try to organise or plan to mitigate this general effect. This chapter is an attempt to identify certain aspects of liberty that might be planned into an extraterrestrial society without compromising the long-term view that what does and does not constitute liberty in space should remain an open discussion and something to be developed by the very people who will eventually live under the conditions beyond Earth.

I should point out from the outset that in this essay I do not discuss the planning of government and how officials are elected, and the necessity or not for democracy. This is an undertaking that merits its own essay. Throughout my discussion, however, it will probably be clear, both overtly and implicitly, that systems of governance in which people are elected and discussion and open debate about the running of a settlement are maximised (whatever mechanism or name we choose to give to this style of governance) are preferable to solving the problems I raise and are more likely to be conducive to the type of vigorous debate needed to resolve the problems in a continuous way.

In this essay, I am concerned with several principal questions: Can we physically engineer a space settlement to be more conducive to liberty? What do we do about the problem of oxygen and maximising liberty when faced with a commodity so prone to tyrannical control? How do we mitigate the worst tendencies of an extraterrestrial surveillance state? How can we plan to maximise economic liberty when the vast spatial scales of space cause economic isolation? Finally, and more generally, I use these examples to ask the question: Can we devise a 'liberty calculus' that allows us to more accurately determine what courses of political, economic and social action are likely to improve the conditions for liberty?

3.2 Engineering Liberty

The first question we might address is whether there are physical aspects of the engineering involved in putting together an extraterrestrial settlement that lend themselves to maximising liberty. If extraterrestrial conditions that enforce enclosure within pressurised spaces tend to put people at the mercy of authorities who control those spaces, can we find ways to liberate people as much as possible?

The ability to move freely through most physical space on Earth (there are obvious trivial examples of places that are not so easy to access, such as the deep oceans, but here I am speaking of places where people live) means that rarely do architects and engineers have in-depth discussions of how liberty is to be maximised using the built environment. There are numerous examples of how architects discuss open spaces, such as public parks and access within structures like high-rise housing blocks of 'new towns',[3] but these deliberations are more focused on creating a good quality of life than avoiding tyranny per se.

In the extremes of space where the outside environment is instantaneously lethal, there are converging problems for liberty. There is the physical problem that the inability to move so freely will curtail people's options for freedom of movement. This is a fact of nature, not a denial of liberty. However, extremes set in place social arrangements, both physical and political, that may cause a truncation of maximum liberty. Individuals and corporations can seek to use the extremity of space as an excuse to limit freedom of movement and control those commodities on which movement depends, such as spacesuits and oxygen-producing machines.[4] Inadvertently, without proper engineering and architectural foresight, the built environment can enhance the extent to which this natural difficulty of freedom of movement transforms itself into an all-embracing suffocating tyranny. Finally, these two factors, human and physical, can subtly generate a social environment that feels compulsive.[5]

The ways in which liberty can be engineered into an extraterrestrial society are legion. Here I would like to focus on two to illustrate the principle. It would merit an entire book in itself to consider how liberty is to be engineered into extraterrestrial societies, but two examples are sufficient to show how it may be possible to plan liberty without the planning itself becoming despotic.

The first element relates to the matter of freedom of movement. For people to feel a sense of control over their private lives and interactions with the polity, it is essential that they are able to move from one place to another without duress and without asking permission. Controlling freedom of movement, particularly across national borders, has been one of the favoured instruments of totalitarian states on Earth for millennia.

One way in which people move from habitat to habitat and settlement to settlement (on the same planetary surface) is by using spacesuits and vehicles. Consider spacesuits for a moment. We can envisage a number of physical characteristics of these objects that would enhance freedom of movement and maximise the ease with

[3]There are innumerable examples of these deliberations. For example, just two are Katz (1993) and Alexander (2009). However, neither of these (or indeed any others I could find) explicitly link urban planning to the historical discussions and precepts of liberty.

[4]The problem is one of using the pursuit of liberty as the very excuse to curtail it. In the terrestrial case, the most in-depth exploration of this problem was made by Berlin (for example, in Berlin (2002)). He suggested a crucial distinction between negative and positive liberty, the former liberty resulting from lack of interference, the latter the extension of the capacity for individuals to realise their goals. The problem Berlin discussed was that positive liberty can become an excuse to establish coercive regimes for the purpose of 'enhancing' people's liberty. If the lethal external environment leads to a 'we know better than you how to secure your liberty against the outside environment', then positive liberty becomes an instrument of tyranny.

[5]Paradoxically, even those in power fall prey to a form of slavery that results from their role as mere providers of dictates. Weil (2002, p. 91) stated that 'As the man of power lives only by his slaves [...] he is never capable, strictly speaking, of willing, but is prey to desires [...] he passes all of a sudden from the feeling of absolute power to that of utter impotence.'

which people can move around a settlement or just take a walk outside to escape for a short trip. The spacesuits should be easily maintained, thus reducing the extent to which they have to be serviced. High servicing demand lends itself to greater power over their use by individuals and corporations; it increases the number of times spacesuits must be taken away to be examined and it increases the fear that they will fail. Spacesuits should be easily mass-produced, thus achieving high redundancy in the number available and the security that at any time an individual chooses to go outside or travel somewhere, they can. These garments should be made of standardised parts to ensure that replacing parts is easy (thus reducing the servicing problem). They should be easily modified to ensure that one size fits many, again increasing the redundancy available for any given individual. The suits should be very comfortable such that wearing the garment for long periods of time does not feel too burdensome or produce a very strong feeling of restriction and entrapment. The characteristics just listed are not necessarily required for the construction of a reliable, functional spacesuit that can be used to do science and exploration.[6] These latter objectives do not mandate a mass-produced spacesuit, just a sufficient number for scientists to launch expeditions. Thus, it is not necessarily the case that by building good spacesuits for space exploration and settlement we end up automatically maximising liberty. We need to design liberty into spacesuits.

Secondly, liberty might be engineered into the habitats themselves. The environment of space may be inherently tyrannical, but it also offers the potential to allow, in Mill's words, 'experiments in living'.[7] These experiments may turn out to be vital for the success of space settlement since we do not know exactly what arrangements of social organisations, particularly at the fine scale, will lead to the most felicitous outcome for the inhabitants. It seems beneficial to give the maximum chance for as many permutations of social organisation to be attempted in order to allow for the successful arrangements to come to the fore.

Quite apart from the flexibility to allow for different social structures, the confines of space habitats by their nature encourage a centrally controlled structure.

[6]There are many scientific papers and texts examining spacesuit design (e.g. Schmidt et al. 2001; Graziosi and Lee 2003; Jordan et al. 2006), but all of them approach the problem from a purely functional perspective. None to date, as far as I know, discuss overtly how spacesuits might be designed to maximise liberty, although in attempting to fathom how they can be made more comfortable, they do this implicitly.

[7]In Chap. 3 of his work, *On Liberty*, Mill (2004) elaborated on the reasons for allowing different 'experiments in living' (p. 57) based on his view that without this freedom, society succumbs to the 'despotism of custom' (p. 70). Indeed, he went a step further and positively encouraged eccentricity (p. 67): 'It does seem, however, that when the opinions of masses of merely average men are everywhere become or becoming the dominant power, the counterpoise and corrective to that tendency would be, the more and more pronounced individuality of those who stand on the higher eminences of thought. It is in these circumstances most especially, that exceptional individuals, instead of being deterred, should be encouraged in acting differently from the mass. In other times there was no advantage in their doing so, unless they acted not only differently, but better. In this age the mere example of non-conformity, the mere refusal to bend the knee to custom, is itself a service. Precisely because the tyranny of opinion is such as to make eccentricity a reproach, it is desirable, in order to break through that tyranny, that people should be eccentric.'

No matter how diverse and experimental social arrangements may be, if the oxygen required by everyone to breathe is controlled at one single point by one single set of individuals or an organisation, then everyone remains at the mercy of these groups, no matter how benevolent they may appear to be. Are there ways in which the architecture of settlements can be engineered to maximise liberty?

It has been proposed that modularity in design could allow for decentralisation of vital commodities.[8] If a set of habitats is fed with water and oxygen from one central location or the food is produced by one food-growth unit, then not only is there a greater chance that these facilities will fall under the influence of the despotically inclined, but these individuals will be positively attracted to controlling the facilities on which the lives of so many people so obviously depend. If the settlement is constructed of habitats fragmented into segments that have their own water, oxygen and food-growth units, there is a less obvious path to control. The control of one segment leaves other segments of habitats potentially free. Of course, this only works if the water, oxygen and food-growth units are not controlled by a single group of individuals or a corporation. This is an important point because there will be little advantage of decentralisation of these assets if they are still controlled by the same people. Indeed, physical decentralisation of vital assets that are still under the control of a single organisation may even increase the effectiveness with which the organisation can target particular groups of people or manipulate the supply of spare parts to crush dissent, etc.

The decentralised architecture can therefore only work if it goes hand-in-hand with social efforts to reduce monopolies and maximise the opportunity for segments of habitats to choose new suppliers if the inhabitants perceive problems with their existing suppliers.[9] This is discussed in the case of oxygen in the next section.

In the event that only one organisation exists to supply water, oxygen and food-growth capabilities, then of the two alternatives, centralised supply and fragmented supply, probably the latter is preferable because central supply runs the risk that the control of the supply threatens everyone.

In the development of 'districts' of habitats, we see the potential for the architectural design of liberty.

3.3 Liberty and the Problem of Oxygen

Of the many commodities required for long-term survival of a space settlement and the establishment of a permanent human presence in space, oxygen provides the most thorough example of something that creates opportunities for the concentration of

[8]Discussed in Cockell (2010).

[9]As Hayek (2006, p. 340) stated, 'the preservation of freedom in the sphere of the mind and of the sprit will depend, in the long run, on the dispersal of the control of the material means and on the continued existence of individuals who are in a position to devote large funds to purposes which seem important to them'.

power and wealth with its potentially dangerous consequences for liberty. Several of its characteristics make it a magnet for the powerful. It is required on a second-to-second basis. Ultimately, of course, we all need food and water as well, but an inherent temporal buffer exists in these latter goods meaning that if we are denied them, we remain alive, perhaps for sufficiently long to attempt revolution or to seize control of the assets that produce them. This is not the case for oxygen, a want of which will cause rapid asphyxiation. There is no time to attempt to seize oxygen-producing machinery if it should be taken over, controlled and used as a coercive instrument. There is no extraterrestrial environment where the oxygen requirement for people can be met without technological intervention. Thus, between the substance that contains the oxygen (for example, water or CO_2 on Mars, or water on the Moon or asteroids) and a breath of fresh air, there is a range of technologies that must extract the substance, process it to generate oxygen, and distribute it in tubes and other apparatus to people and processes to repair all of the machinery that is required for this complex interconnected process.

That technology is required to gather and distribute oxygen is an ineluctable fact of every known extraterrestrial environment. Therefore, we are left with the conclusion that people *will* control processes between substances that contain oxygen atoms and the people that breathe it. The question then becomes: Are there means by which we can minimise dictatorial control over the processes that can be predicted and implemented prior to the construction of extraterrestrial settlements?

There are a number of social and economic means by which such a state of affairs may be avoided and I will touch on just a few of these here. First, we might implement legal agreements and regulations that minimise monopolies. If the number of producers of oxygen can be increased above one, then there will be greater choice for people to decide on providers, and even if those choices are limited, it will be less likely that an entire settlement can fall prey to a single corporation.[10] The avoidance of monopolies may be difficult in the early years of a settlement because the number of corporations building a settlement may be limited. The establishment of control over oxygen in the early years of a settlement will be difficult to break later if a corporation refuses to yield economic control over oxygen production. There is little to be said of this possibility other than that it has a

[10]Of course, this is underpinned by the assumption that if the legal structures are in place, there is a profit to be made in oxygen production. We cannot easily predict this at the current time (see chapter by Adam Stevens, this volume). Even Smith (1776, Book V, Chap. 1) recognised that some processes may not make a profit but are essential to society. In these, he included 'public works, which, though they may be in the highest degree advantageous to society, are, however, of such a nature that the profit could never repay the expense to any individual or small number of individuals'. If it eventually turns out that oxygen falls within this category, then people will still need to find ways to minimise monopolies in the public provision of this gas.

good chance of leading to tyranny.[11] The people involved in designing a settlement and the corporations and/or state entities involved in building it, should, if they care for the long-term liberty of the people there, do everything in their power to multiply the number of oxygen-producing entities and minimise monopolies.[12]

Little advantage is to be gained if multiple oxygen-producing entities all collude to control oxygen, thereby giving the illusion of competition under a regime of coordinated price fixing and technological control. This situation will itself lead to nothing better than a monopoly. There are a number of ways in which the probability of such an outcome can be minimised.

There can be no substitute for open and free access to information.[13] Systems that allow people to see who is on the boards and committees of corporations have a good chance of revealing collusion. These systems of open corporate information can never be completely successful. Even if there is no overlap in board or committee memberships of organisations, these institutions may still collude. An additional measure is to create independent boards, such as the Extraterrestrial Oxygen Board, comprised of elected members of the settlement, who are tasked with discussion and open reporting (perhaps on a six-monthly basis) on the state of the oxygen industry and evidence that competition is open and fair.[14] Yet, again, these systems can never be completely failsafe. These types of committees themselves become the focus of jealous corporations seeking to control the mechanisms of independent oversight and reporting. Like any similar committees on Earth, by requiring such committees to openly report in a way that can be scrutinised by the

[11]However, the hope that an initial situation of authoritarianism can successfully transition to a more open society is not in vain. There are countless examples of this transition on Earth and a great deal of literature exploring how this can be done, much of which could be applied to the extraterrestrial case. For example, O'Donnell and Schmitter (1991) discussed the process of increasing democratisation during the breaking down of autocracies. They offer specific examples of methods (p. 49): 'Although we cannot provide hard data to prove it, our personal experience in having lived through several of these moments indicates that the catalyst in this transformation comes first from gestures from exemplary individuals.' There are other methods that can be identified and they merit a systematic search.

[12]There is a definitional point here. We do not know precisely what sort of apparatus of governance will evolve in different environments and it would be an error to assume that only when something becomes a 'state' can it be coercive. Any type of governance can be tyrannical, as von Mises (1944, p. 46) recognised: 'But not every apparatus of compulsion and coercion is called a state. Only one which is powerful enough to maintain its existence, for some time at least, by its own force is commonly called a state.' Coercive control of an extraterrestrial settlement is possible at any level.

[13]This was a point recognised by Bury (1952, p. 225): 'It requires positive efforts and a sustained national faith to make sure that independent individual thinking is encouraged and facilitated and that adequate relevant information is available wherever policies are made, and that such policies and the responsibility for them have the requisite publicity.'

[14]The problem is an ancient one. As Russell stated (2004, p. 81), 'The exercise of power, if it is to be something better than the infliction of wanton torture, must be hedged round by safeguards of law and custom, permitted only after due deliberation and entrusted to men who are closely supervised in the interests of those who are subjected to them.' The latter proviso is particularly important in the case of extraterrestrial oxygen production.

public, there is some chance that cabal and corruption can be minimised and the environment improved for fair competition.

These conditions may be difficult to achieve in the early years of a settlement as the number of people will be small. The prospect of oversight committees and other instruments is better realised in large populations. There is little doubt that in the very early years of a settlement, a certain faith in the behaviour of people is required to ensure that the initial opportunities that will always exist in a small settlement for monopolisation of the means of production does not rigidify into a systematic tyranny as the settlement develops. The implementation of oversight committees and transparent discussion and reporting from the outset can go a long way to mitigating autocratic tendencies.[15]

Aside from the social and governance aspects, as with spacesuits and habitats, there is much that can be done in the engineering of oxygen-producing machinery to minimise the tendency to inflexible central control. Oxygen-producing machines can be made as simple as possible using cheap spare parts that any individual has the financial means to acquire. They should be built with simplicity and flexibility so that even if a spare part is not available, it may be possible for an individual to improvise with other materials. Oxygen-producing machines should be designed to incorporate as much of the process as possible that is required to go from substrate-containing oxygen atoms to breathable oxygen. This engineering approach minimises the number of steps between substrate and breathable oxygen, thus reducing the number of links in the chain that offer themselves up to oversight from someone else. An oxygen-producing machine that has all these steps integrated into it increases the possibility that all an individual has to do when faced with a possible lack of oxygen is find the suitable substrate to produce it (which could be atmosphere on Mars or ice on asteroids) and they can, independently of others, produce sufficient oxygen to breathe. This approach to design is tantamount to a type of oxygen independence.

The substrates from which oxygen might be produced are also a matter for the planning of liberty, particularly where it might be produced from the atmosphere. For example, on Mars, oxygen could be produced from several sources. It could be produced chemically from the atmosphere by 'cracking' carbon dioxide. It might be produced biologically from the atmosphere by pressurising the atmosphere and supplying it to plants or it could be produced by melting and cleaning water in permafrost and electrolysing it.[16] At the equator and mid-latitudes on Mars, the greatest extent of liberty can be secured by extracting oxygen from the atmosphere

[15]Brenkert (1991, p. 190) nicely summarised three conditions that apply to the extraterrestrial case in the matter of open discussion of policies: '… the scope of participation must be at least threefold: (a) citizens must be able to participate in decisions whereby the basic directions of their society are determined; (b) they must be able to participate in determining who fill the posts through which the directions society takes are determined, and (c) they must be able to participate in creating or reinforcing social and political values and institutional practices that determine which issues will come within the scope of the political process and public consideration'.

[16]Just three of these discussions include Sridhar and Vaniman (1997), Mitchell (1994) and Sridhar et al. (2000). For the lunar case, an example is Taylor and Carrier (1993).

because, as on Earth, it is difficult for authorities to deny people gas from the atmosphere. If an oxygen-producing machine is of sufficient good design and simplicity to produce oxygen from the atmosphere, then individuals have the potential to produce oxygen. If oxygen must be produced from permafrost, then the digging, cleaning and electrolytic treatment of the water requires considerable effort that may be beyond the power of the individual or at the least, puts them at the mercy of the collective effort required to get at this resource, which could be curtailed by a determined authority. On Mars then, it seems clear that oxygen-producing machinery that can produce oxygen from the atmosphere is a better liberty-protecting alternative than systems that require it from the ground. Unfortunately, on the Moon and asteroids, water must be gathered from ice and so the practical process required to remove it from ice deposits is an unavoidable requirement for breathing, raising the possibility that on Mars, at least with respect to oxygen production, liberty may be easier to protect.

3.4 Liberty: Controlling the Surveillance State

The lethality of the extraterrestrial environment forces those managing the settlements to take a thorough and responsible approach to ensuring the safety of the occupants. Not only will those who govern and oversee the engineering of a settlement be required to provide adequate safety checks, but the people themselves will demand this.

The greatest safety threats come from the possibility of depressurisation, which has the potential to kill large numbers of people rapidly. Failures in food and water production certainly will be serious, but they do not impose the possibility of instantaneous catastrophe. How is the integrity of the habitats and other infrastructure required for the settlement, including spacesuits, rovers and other apparatus, to be maintained? The obvious answer to this is either regular checks by human or robotic operators or a systematic network of cameras and sensors, both of which will allow for ongoing checks on infrastructure.

At once we see the potential for tyranny. Driven by a fear that lack of safety checks and oversight will lead to catastrophe, a fear reinforced by collective demands for efficient oversight, a network of sensors, cameras and officials responsible for making checks leads to a structure that is both attractive and amenable to control by powerful elites.[17] This network offers the possibility for

[17]Although one should be under no doubt that some type of bureaucracy is desirable to prevent catastrophic failure. The question is the extent to which it is allowed to dominate people's public and private lives. As von Mises (2007, p. 14) summarised, 'But people are unfair in indicting the individual bureaucrat for the vices of the system…The system is bad, not its subordinate handymen… What people resent is not bureaucratism as such, but the intrusion of bureaucracy into all spheres of human life and activity. It is a misnomer to label the fight for freedom and democracy a fight against bureaucracy.'

control in a number of ways: the officials become servants to special interests and authorities, becoming overbearing, petty and demanding to the general populace both to demonstrate their allegiance to the central authorities and to exercise a natural desire to achieve control and power over other people as an indication of social importance[18]; the sensors and cameras become a technological instrument of control allowing for remote monitoring of people and social groupings under the pretext of checking physical infrastructure; safety checks become a convenient way of targeting specific people or groups to remind them who is in control and to worry them into submission by convincing them that their lives are threatened by some defective engineering; the possibility of catastrophe at any moment becomes a convenient excuse for a generally coercive frame of mind and a pre-emptive approach to policing—it is better to detain someone who might be intent on criminal damage now and find them innocent later than to allow them to run free with the potentially disastrous consequences for everyone; and the safety culture becomes a mechanism to develop a culture of conformism, where the radical individual is a threat to the safety of everyone and hence, without formal methods, an individual becomes subject to enormous peer pressure to conform, to check that their actions are acceptable to their fellow inhabitants and to reign in countervailing opinions that are likely to attract the attention of the authorities.[19]

The not insubstantial list elaborated above provides just some of the unspoken and deliberate means by which the deadly space environment and the safety checks that it requires leads to an all-pervasive surveillance state-dominated control culture supported by extensive technological capacity for monitoring people.

If we accept that the extraterrestrial environment has the inevitable potential to drive society in this direction, then the obvious question for those seeking to maximise liberty is: What can be done to mitigate this tendency? There are ways in which the worst excesses of the surveillance state can be reduced. These methods are both technological and social.

On the technological side, one obvious way to limit the control of the authorities is to minimise the pervasive presence of technology that records video and sound. Many of the potential problems with infrastructure that could lead to depressurisation, for instance, may be manifest as stresses in structures or changes in structural integrity. These characteristics can be monitored without video or sound

[18]Skinner (1998, p. 119) observed that 'One side argues that the state can hope to redeem this pledge simply by ensuring that its citizens do not suffer any unjust or unnecessary interference in pursuit of their chosen goals. But the other side maintains that this can never be sufficient, since it will always be necessary for the state to ensure at the same time that its citizens do not fall into a condition of avoidable dependence on the good will of others. The state has a duty not merely to liberate its citizens from such personal exploitation and dependence, but to prevent its own agents, dressed in a little brief authority, from behaving arbitrarily in the course of imposing the rules that govern our common life.' These observations are equally pertinent to the extraterrestrial case, even if the 'state' is nothing more than a few individuals running an extraterrestrial settlement.

[19]The point is that it does not require dictates to create a coercive culture. The mere presence of peer pressure is sufficient to generate a repressive society, a point recognised by Mill (2004). It is discussed for the extraterrestrial case in Cockell (2008).

recording, thus allowing for a large-scale sensor network without necessarily giving authorities the opportunity to use these networks to control people. Of course, cameras have other uses, not least to prevent criminality. Where these devices are installed, they might be decentralised, thus allowing recording and monitoring to be undertaken in sections of habitats by individuals assigned to prevent criminality, without the cameras themselves being physically linked into a central network of cameras that could be used by a single authority.

There can be no hesitation in saying that the best way to minimise the use of surveillance as an instrument of control is to generate a culture of open information. There are a number of mechanisms by which this can be achieved. The technical details of surveillance structures can be made public. Although this risks making available information that could be used by criminals intent on causing damage, I will assert here that overall the gains to be had in enhancing liberty by giving people the opportunity to become aware of surveillance networks and monitor their extent and use are greater than the potential use of this information for criminality.

Social mechanisms of enhancing the chances of holding back an excessively ambitious surveillance culture include a free and open press and mechanisms to challenge safety edicts. The former is an effective brake by allowing people to express concerns about aspects of oversight, to discuss the efficacy and usefulness of health and safety directives and generally to engage in open dialogue about the relationship between authority and its power to observe.[20] The latter requires the formation of publicly electable bodies that ensure that the implementation of surveillance and health and safety checks is both necessary and reasonable and to act as a conduit between the wider public and the people running the settlement.

A free and open press can be accomplished by something as trivial as setting up newspapers. Like the amateur papers established on polar bases on Earth, small, freely distributed newspapers (the *Lunar Times*, the *Martian Chronicle*, etc.) can do much to promulgate opinions, sharpen discussion and encourage an open and vigorous culture of dissent, satire and the free exchange of ideas.

As with so many of these proposed strategies, there is clearly a link between the scale of the settlement and their effective implementation. For small settlements—as is the case for most of them in their early establishment—there is less chance for the separation of powers (for example, between the people that run the settlement and implement surveillance and the people charged with independently overseeing this). The possibility always exists that this paucity of choice in people to carry out these different functions will lead to a cabal that is difficult to break even when the

[20]It is clear that for this to be realised, people must be technically and scientifically educated to be able to assess all of the systems that surround them as necessary or unnecessary coercion. The extreme case is children, for example Fried's point (2007, p. 172) '…that we are born ignorant, unreasoning, and dependent. It should not be the excuse and occasion for stamping out our capacity for liberty at the very beginning. The dilemma is a challenge to the ingenuity and liberality of the liberal spirit to do the best it can.' This point is equally applicable to the education of adults, which should be an ongoing concern for the construction of open extraterrestrial societies.

population expands, particularly if the initial settlers have entrenched special interests that they and subsequent generations seek to protect. As I have noted before, there is very little that can be said of this. If people allow this to happen and those who establish these outposts are not motivated to lead it towards an open culture of liberty, then tyranny is assured. Either the founders must see the value of freedom to enhance the chances of survival of the very settlement they have worked hard to construct or new generations and occupants must consider removing the existing structures and replacing them by more independent political and economic structures that enhance the chances of an open culture.

3.5 Planning Economic Liberty

There are irrefragable characteristics of the extraterrestrial environment that impinge on the economic possibilities beyond Earth. No matter how much we might be enthusiastic about the possibility of access to energy and vast quantities of material resources to both drive the settlement of space and supply Earth,[21] these options will be limited by several problems, all of which stem from the physical reality that the vast scales of space will make communities remote from each other. There is therefore a tendency to economic isolation.

Economic isolation can cause a number of deleterious knock-on effects. It reduces the potential for new goods to come into a settlement, thus limiting choice to goods only produced in the domestic market. The lack of ease with which corporations can trade in isolated settlements reduces external competition, which is likely to result in false price structures, distorted markets and not least, the possibility of monopolies or oligopolies with price fixing.[22] The free movement of workers will also be stymied, increasing the chances of economic tyranny over workers, reducing the workers' chances that they can voluntarily remove their labour from a given entity by moving elsewhere and putting them at the mercy of authoritarian managers and corporate decisions on their salaries and quality of life

[21]For example, this is explored by Lewis (1997).

[22]Corporations (and commerce in general) may have the beneficial effect of stabilising a settlement since it is in their interests to have calm and consistency for the running of a corporation. In mitigating the chances for revolution, de Tocqueville (1998, p. 323) recognised the following: 'I know of nothing more opposite to revolutionary attitudes than commercial ones. Commerce is naturally adverse to all the violent passions; it loves to temporise, takes delight in compromise, and studiously avoids irritation. It is patient, insinuating, flexible, and never has recourse to extreme measures until obliged by the most absolute necessity.' Although this may be true, and it may well generate some predictability in an extraterrestrial case, of course, the countervailing effects can also exist, namely that any individual or organisation that seeks a new way forward or to question existing arrangements, particularly in a settlement with strong corporate presence, may well be crushed by a corporation seeking to 'avoid irritation'.

that they will find difficult to avoid or escape.[23] Put simply, there is a greater chance of Victorian style slavery, even in the presence of unions or other groups established to represent workers.[24]

The economic isolation is likely to fuel the flames of several political problems. The chances of autocracy are enhanced as the corporations yield great control over the economic scene which collapses into collusion and corruption in the political sphere. The strong peer pressure and control that results from the lack of freedom of movement of workers enhances a general culture of conformity which feeds political control over the settlement and the destruction of individualism and dissent. Put simply, extreme extraterrestrial isolation degrades liberty.

How are these effects to be minimised? One clear way to do this, particularly within the solar system, is to maximise the possibility of reliable and widespread trade routes. The greater the number and rapidity of these routes, the greater the possibility of effective external competition and movement of people with all of its side effects of discouraging an isolated repressive environment. This in itself can be achieved by the availability of cheap, mass-produced spaceships for moving people, products and raw materials across the solar system. Most people consider these developments to be technically desirable as they increase the potential success of mining and other commercial operations, but they are equally important for maximising liberty. Ironically, reliable trade routes and rapid movement of goods and people are as important for commercial development as they are for mitigating the potential tyranny that results from the success of the very commercial enterprises using these trade routes.

Continuous vast trade routes are not assured and another parallel way to mitigate the problems elaborated here is to maximise domestic competition and choice, thus reducing the chances of distorted markets and price fixing even in the condition of isolation. This is to be achieved by settlement constitutions that maximise the

[23]They must, however, be aware of their subjugation and this requires effective education. As Muller (1970, p. 79) recognised, 'Primitives may consider themselves free, or more precisely not feel unfree, because they are unconscious of the constraints we perceive [...] Although we cannot be sure of the state of mind of the illiterate peasant masses throughout history, their usual passivity suggests that they generally took their subjection for granted.' Although it is difficult to imagine this same lack of awareness in anyone who is working in a space settlement, the point is that the extremity of the extraterrestrial conditions can be used to convince people that they are not subjugated, but instead experiencing the maximum level of freedom possible against the total destructiveness of the external environment, and therefore they should acquiesce to decisions. Individuals and societies must guard against this. Boas (1942, p. 55) equally recognised this: 'We believe that we have such freedom and are not aware of our own limitations [...] In this sense we may say that absolute freedom does not exist [...] We are free in so far as the limitations to our culture do not oppress us; we are unfree when we become conscious of these limitations.'

[24]Patterson (1991, p. 99) emphasised the link between slavery and freedom in ancient Greece, where slavery was accepted since the rulers were kin in a world facing barbarians. If we replace barbarians with the lethality of the extraterrestrial environment, then we can see how economic slavery may even be passively accepted by a population. Patterson observed that 'The demos accepted the rulership of the traditional ruling class because they saw its members as kinsmen, kith and kin against a world of unfree barbarians.'

freedom with which people can set up and establish commercial entities and allow for external entities to freely enter into settlement markets (essentially what amounts to interplanetary free trade). This can only be achieved with an open, non-venal political environment that encourages open discussion and transparency of information. As on Earth, economic freedom and expansion in space is inextricably linked to political freedom.[25]

3.6 Liberating Creativity

Despite the predictions of the conformity-causing extremes of space, the environment of outer space offers a release from previous mores and social organisation. New forms of art, music and other cultures offer the potential to encourage an environment of creativity as an antidote to the environment.[26] This is the case as much in settlements as on much larger spaceships, even those heading to interstellar destinations.

Many forms of art and human creativity do not require free movement. They include painting, music, philosophy and other forms of art, so in many ways they lend themselves to expression in space in ways no more restricted than on Earth. This implies that no special physical planning is required for them to flourish. However, what is required is the political environment that allows them to be exercised. Freedom of expression must be incorporated into laws and constitutions in explicit recognition of this opportunity.

We can therefore plan for an environment in which freedom of expression in all its forms is encouraged.[27] Not only will this offer the opportunity for individuals to realise their potential, but in practising this freedom, the collective culture of freedom of thought will be encouraged and the tendency to conformity reduced.

It would seem that many of the lessons learned on Earth regarding the separation of the state (the settlement committees and structures of oversight in small settlements) and these artistic activities can be applied in the extraterrestrial environment. The obvious example is religion.[28] The encouragement of freedom of expression brings with it the opportunity for small groups of determined individuals in a

[25]This is explored particularly forcefully by Milton Friedman (1995). A more extreme version, bordering on anarcho-capitalist, was also explored by another Friedman—Friedman (1962).

[26]The notion that space offers an intellectual release has been explored previously. Crawford (2014) suggested that a programme of interstellar exploration could be undertaken for the purposes of avoiding intellectual stagnation, including in science, art and philosophy. Beattie (2014) used experiences in analogue environments on Earth to show how art practices can be used to enrich the extraterrestrial experience. Cockell (2008) suggested that the extremity of the extraterrestrial environment will actively drive people to seek intellectual release through novel creativity.

[27]See the chapters by Annalea Beattie in this volume and in Cockell (2014) for a more thorough discussion on the link between art, the extraterrestrial environment and liberty.

[28]This was partly explored for the extraterrestrial case by Cockell (2008, 2009).

settlement to dominate over quieter factions. Thus, freedom of expression can quickly result in the monopolisation of culture and therefore ultimately real structures of control by small groups or special interests with a particularly strong agenda. This is more likely to happen in small settlements where it is physically easier to dominate. The apparatus of management of a settlement should remain impartial to the emergence of different forms of art and should encourage freedom of expression, for example through a free press to allow for diversity of opinions to be overtly experienced and to allow for concerns from members of settlements to be expressed.

3.7 The Ultimate Planning of Freedom

Finally, it is worth mentioning the possibility of the ultimate act of planning liberty: engineering the environment of a planetary body to make freedom of movement greater. For many planetary bodies that lack an atmosphere, this possibility does not exist. However, for a body marginally close to being habitable, there is the possibility to engineer the atmosphere. In our solar system, one example is Mars.[29]

The introduction of chlorofluorocarbons (CFCs) into the Martian atmosphere is calculated to cause the release of carbon dioxide, thus increasing surface temperatures and atmospheric pressure. Once conditions on the planet become suitable for liquid water, photosynthetic organisms could be introduced to produce oxygen.

From a human point of view, terraforming would allow people to walk on the surface of Mars using something akin to scuba equipment once surface pressures become high enough. The ultimate goal would be atmospheric oxygen concentrations allowing for breathing without technological support. However, even if this goal is not reached, terraforming can be regarded as an instrument of liberty and the ultimate attempt to engineer freedom.

If we accept that the lethal atmosphere (the lack of breathable atmosphere) on planetary bodies is one of the key drivers of tyranny through the way it curtails freedom of movement and offers the potential for control of oxygen supply, then any technology that mitigates this will make significant strides in increasing freedom.

Nevertheless, there are limitations. First of all, the technological difficulties of terraforming are enormous and I will not elaborate on them here. They could take many centuries or millennia to overcome. Secondly, terraforming does not solve other factors that lead to extraterrestrial tyranny. It does not change the reality of isolation (economic and political isolation) on a planetary body. Even in conditions where people can roam freely on a planetary surface in a breathable atmosphere, the problems related to economic isolation still have to be solved.

[29]In-depth discussions on this technological possibility are to be found in McKay et al. (1991) and Fogg (1995).

3.8 A Liberty Calculus

This is an apposite time to return to the beginning of this essay and the concern raised about the potentially tyrannical consequences of 'planning' liberty. How are we to implement some of the ideas expressed in this essay and others besides without doing the very thing we wish to avoid, namely generating a social system where the good intentioned end up controlling erroneous directions of liberty in space?

One way in which any schemes can be assessed as to whether they offer the potential to maximise liberty or whether they represent a misguided attempt to manipulate the direction freedom takes is to have an informed discussion on the potentially tyrannical and liberty-maximising paths that different choices might take. We might call such a system a 'liberty calculus'. Consider, for example, the design of spacesuits (as discussed earlier in the chapter) that are engineered to maximise the ease with which they can be used to move around between habitats and outside a settlement. We could ask ourselves the question: What are the relative contributions of two different design trajectories to liberty and tyranny and therefore which one is best? We could then draw up a matrix to consider this question.

In Table 3.1 I have attempted such an exercise. We ask ourselves what opportunities each design path offers to liberty-seekers and tyrants and then attempt to assess which one is preferable to liberty. We could even attempt some sort of crude quantification by giving points to the extent to which different forms of liberty are attained and negative points for the erosion of liberty, although one should be cautious for such an assessment not to deteriorate into an excruciatingly detailed Benthamist utilitarian calculus.

Table 1 Planning liberty

	Spacesuit is mass-produced, easily repaired, and easily donned and doffed	Spacesuit is merely designed for use in science and engineering without maximising movement and availability
Beneficial consequences for liberty	• Increased freedom of movement; ability to go from one habitat to another with ease; ability to travel from the settlement on excursions easily; ability to get spare parts easily. Increases general feeling of freedom to move around. Reduces ability of central authority to control movement	• No good consequences for liberty. The effect is to restrict movement and thus the culture that emerges from free movement
Beneficial consequences for tyranny	• Increases ease with which people acting on behalf of dictatorial agencies can move around	• Enhances capacity to control population, restrict movement and coerce individuals. Maintains spacesuit production under control of limited individuals

A simple qualitative matrix to aid in the decision of whether it is a good idea for the maximisation of liberty to design spacesuits that are mass-produced, easily repaired and easily used by people

In Table 3.1 we can see that the building of spacesuits for freedom of movement generates advantages for both the liberty of the population and for ironhanded authorities. However, the building of spacesuits that restricts freedom of movement generates no advantage to the population in terms of their liberty but many advantages to a dictator. The conclusion we arrive at is that designing spacesuits that maximise freedom of movement is objectively a good thing to do for liberty. In this case, we can reliably plan for liberty.

The point I want to make is that for some proposed social or engineering projects, it may be possible to qualitatively perform a calculus on the merits and demerits of various schemes for enhancing liberty in the extraterrestrial environment. It is not necessarily the case that the idea of 'planning' for liberty is one fraught with unavoidable conflicts and disagreements on the definition of liberty. In some ways, it is the extremity of the space environment that makes the consequences of tyranny so great that the polarised nature of some of these decisions makes their outcome, and therefore their planning, easier to divine compared to Earth. In cases where the outcome for liberty between two or more paths is uncertain, we most likely have an example of a situation where we should not attempt to pre-plan, but rather leave the process to an organic set of decisions arising from the polity. An example I can think of is decisions on the types of food that are to be grown in an extraterrestrial environment. Apart from the obvious extreme example of choosing food that leaves people so nutritionally deficient that they rely on a tyranny to keep them alive, in all other situations we can hardly imagine 'planning' liberty into food choices. This is something to be discussed and subject to change over time, not least by changing fashions and social preferences.

3.9 Conclusion

The environment of outer space has a tendency to encourage totalitarianism. However, there are ways to minimise these tendencies. Planning for liberty is a concept wrought with dangers. It can easily be used as an excuse for control, compulsion and coercion under the pretext that the efforts are well-meaning attempts to improve people's liberty. In this essay, I have suggested some of the ways in which liberty might be planned and how it may be possible to carry out analyses that allow one to more objectively discern which practices will advance liberty, and can therefore be planned, and which practices should be left to an emerging consensus.

References

Alexander, A. (2009). *Britain's new towns: Garden cities to sustainable communities*. New York: Routledge.

Beattie, A. (2014). Extraterrestrial liberty and creative practice: A firsthand experience of an imagined future. In C. Cockell (Ed.), *The Meaning of liberty beyond Earth* (pp. 103–116). Berlin: Springer.

Berlin, I. (2002). *Freedom and its betrayal: Six enemies of human liberty*. Princeton: Princeton University Press.

Boas, F. (1942). Liberty among primitive people. In R. Anshen (Ed.), *Freedom. Its meaning* (pp. 50–55). London: Simon Shand.

Brenkert, G. (1991). *Political freedom*. New York: Routledge.

Bury, J. (1952). *A history of freedom of thought*. New York: Idebate Press.

Cockell, C. (2008). An essay on extraterrestrial liberty. *Journal of the British Interplanetary Society, 61*, 255–275.

Cockell, C. (2009). Liberty and the limits to the extraterrestrial state. *Journal of the British Interplanetary Society, 62*, 139–157.

Cockell, C. (2010). Essay on the causes and consequences of extraterrestrial tyranny. *Journal of the British Interplanetary Society, 63*, 15–37.

Cockell, C. (Ed.). (2014). *The meaning of liberty beyond Earth*. Heidelberg: Springer.

Crawford, I. (2014). Avoiding intellectual stagnation: The starship as an expander of minds. *Journal of the British Interplanetary Society, 67*, 253–257.

de Condorcet, A.-N. (1979). *Sketch for a historical picture of the progress of the human mind*. Westport, CT: Hyperion Press.

de Tocqueville, A. (1998). *Democracy in America*. England: Wordsworth.

Fogg, M. (1995). *Terraforming: Engineering planetary environments*. Warrendale, PA: SAE International.

Fried, C. (2007). *Modern liberty*. New York: W.W. Norton and Co.

Friedman, D. (1995). *The machinery of freedom*. Chicago: Open Court.

Friedman, M. (1962). *Capitalism and freedom*. Chicago: University of Chicago Press.

Graziosi, D., & Lee, R. (2003). I-Suit advanced spacesuit design improvements and performance testing. *SAE technical paper* 2003-01-2443, doi:10.4271/2003-01-2443.

Hayek, F. (2006). *The constitution of liberty*. London: Routledge.

Jordan, N., Saleh, J., & Newman, D. (2006). The extravehicular mobility unit: A review of environment, requirements, and design changes in the US spacesuit. *Acta Astronautica, 59*, 1135–1145.

Katz, P. (1993). *The new urbanism. Toward an architecture of community*. New York: McGraw-Hill.

Lewis, J. (1997). *Mining the sky: Untold riches from the asteroids, comets and planets*. New York: Basic Books.

McKay, C., Toon, O., & Kasting, J. (1991). Making Mars habitable. *Nature, 352*, 489–496.

Mill, J. S. (2004). *On liberty and other writings*. Cambridge: Cambridge University Press.

Millar, J. (2006). *The origin of the distinction of ranks*. Indianapolis: Liberty Fund.

Mitchell, C. (1994). Bioregenerative life-support systems. *American Journal of Clinical Nutrition, 60*, 8205–8245.

Muller, H. (1970). Freedom as the ability to choose and carry out purposes. In R. Dewey & J. Gould (Eds.), *Freedom: Its history, nature and varieties* (pp. 75–84). London: Macmillan.

O'Donnell, G., & Schmitter, P. (1991). *Transitions from authoritarian rule: Tentative conclusions about uncertain democracies*. Baltimore: The John Hopkins University Press.

Patterson, O. (1991). *Freedom: Freedom in the making of western culture*. London: Basic Books.

Russell, B. (2004). *Power*. New York: Routledge.

Skinner, Q. (1998). *Liberty before liberalism*. Cambridge: Cambridge University Press.

Schmidt, P., Newman, D., & Hodgson, E. (2001). Modeling space suit mobility: Applications to design and operations. *SAE technical paper* 2001-01-2162, doi:10.4271/2001-01-2162.

Smith, A. (1776). *The wealth of nations*. London: W. Strahan and T. Cadell.

Sridhar, K., & Vaniman, B. (1997). Oxygen production on Mars using solid oxide electrolysis. *Solid State Ionics, 93*, 321–328.

Sridhar, K., Finn, J., & Kliss, M. H. (2000). In-situ resource utilization technologies for Mars life support systems. *Advances in Space Research, 25*, 249–255.

Taylor, L., & Carrier, D. (1993). Oxygen production on the moon: An overview and evaluation. In J. Lewis, M. Matthews, & M. Guerrieri (Eds.), *Resources of near-Earth space* (pp. 69–108). Tucson: University of Arizona Press.

von Mises, L. (1944). *Omnipotent government*. New Haven: Yale University Press.

von Mises, L. (2007). *Bureaucracy*. Indianapolis: Liberty Fund.

Weil, S. (2002). *Oppression and liberty*. New York: Routledge.

Chapter 4
Toward a Sustainable Land Use Policy for Mars

Jacob Haqq-Misra

Abstract Looming prospects for Mars colonization, along with longer-term visions of broader space colonization, open up new territory for human settlement and raise new questions about how unsettled land should be allocated. This scenario has many analogs in human history when settlers in all parts of the world left their home in search of new land for hunting, farming, or exploration. Today no human on Earth can pick up and move to unclaimed land, but space colonization could make this possible again. Here I discuss some historical attitudes toward land appropriation and the shortcomings of these ideas when applied to the human colonization of Mars. This suggests the need for new policy solutions to guide the colonization of Mars prior to the arrival of the first humans.

Keywords Mars exploration · Space colonization · Environmental ethics · Land use policy · Outer space treaty

4.1 A Historical Perspective of Land Use on Earth

Following the decay of the Roman Empire, Germanic tribes successfully managed to assert their presence and displace other tribes in central Europe. These Germanic tribes included a nobility class, a class of slaves, a small class of serfs, and a large number of freemen. As freemen, these people had the right to settle their accounts at any time and move to a new geographic location at will. At this time in Europe, most of the land surrounding these clan villages was unclaimed and not under the sovereign control of any nation. The idea of land ownership did not exist in its modern sense, but freemen typically laid claim to as much land as necessary to support their grazing animals, farmstead, family, and slaves (Hargrove 1989).

J. Haqq-Misra (✉)
Blue Marble Space Institute of Science, 1200 Westlake Ave N Suite 1006,
Seattle, WA 98109, USA
e-mail: jacob@bmsis.org

C.S. Cockell (ed.), *Human Governance Beyond Earth*,
Space and Society, DOI 10.1007/978-3-319-18063-2_4

Expansion of these Germanic tribes (as well as their later Saxon counterparts in England) continued as sons left their family farmstead to claim land on the periphery of the clan village, and it was only when unoccupied borderlands became unavailable that inheritance along selected family lines, usually the firstborn, became commonplace. This right of primogeniture ultimately led to a feudal system of labor that dominated medieval Europe and forced most freemen into an indentured life of serfdom.

This European heritage of land expansion by freemen also inspired American attitudes toward land ownership. Early American interpretation of this tradition defined the owner of land as the people who worked to cultivate, industrialize, or otherwise improve the land from its wild state. The idea that the value of land is inherently based upon its improvement by human labor was articulated by John Locke in his *Second Treatise on Government*:

> And therefore he that incloses land, and has a greater plenty of the conveniences of life from ten acres, than he could have from an hundred left to nature, may truly be said to give ninety acres to mankind: for his labour now supplies him with provisions out of ten acres, which were but the product of an hundred lying in common. I have here rated the improved land very low, in making its product but as ten to one, when it is much nearer an hundred to one… (Locke 1690).

The implications of this land use philosophy even inspired Thomas Jefferson to consider the English kings claim to Western lands as an infringement of the basic rights of Americans to claim and work land for themselves (Hargrove 1989). In addition to its role in instigating the American Revolution, this attitude toward land use was fully manifested when a series of federal Homestead Acts opened undeveloped Western territory to any settlers willing to move and work the land.

Homestead Acts in the United States were passed in 1862, 1909, and 1916 to encourage American settlers to claim undeveloped federal territory by cultivating the land and filing a deed of title. Homesteading follows in tradition with the rights of freemen, except that American laws allow for land to be owned (whereas the concept of land ownership did not necessarily exist in pre-Medieval Europe). Likewise, as with the Germanic and Saxon analogy, American homesteading could only continue as long as undeveloped land remained. Ultimately the Federal Land Policy and Management Act of 1976 eliminated all homesteading programs in the United States, with a 10 year provision for the remote state of Alaska, and on October 21, 1986, homesteading officially ended in Alaska—the aptly named last frontier—making it impossible for a person to pick up and move to unclaimed land. Aside from a remote part of Antarctica (most of Marie Byrd Land) and a border region between Egypt and Sudan (Bir Tawil), all land on Earth today has been claimed by sovereign nations.

An obvious deficiency in the freeman attitude toward land use is the assumption that land is unlimited. John Locke even makes the erroneous assumption

> …that every man should have as much [land] as he could make use of…since there is land enough in the world to suffice double the inhabitants (Locke 1690).

Although Locke may have underestimated the rapid growth of human population over the centuries that followed him, he also may not have considered that space exploration would provide a new frontier for human colonization. Homesteading, either from lack of law or by provision of law, has been a way of life throughout most of human history. Perhaps the era of space colonization will once again renew opportunities for homesteading. Yet even space resources are finite, and unrestricted homesteading for space colonization will eventually face the same limits to growth as the Germanic freemen or American settlers. Any such policy that follows the tradition of homesteading will relegate control of space resources to those who are first to arrive.

Another critique of Locke's philosophy and its justification of homesteading is the philosophical idea that land itself might be valuable, independent of cultivation or improvement by humans. Lockes position, carried out to the extreme, is reminiscent of certain biblical or Aristotelian views that Earth and its resources exist exclusively for the benefit of human beings, a view that has been widely criticized but still has modern adherents (White 1967; Hardin 1968; Haqq-Misra 2007). However, the idea that land could be intrinsically valuable, apart from its utility to humans, began to emerge during the environmental movement of the 1950–1970s and led to the formulation of environmental ethics as a discipline (Rolston 1988; Callicott 1989; Hargrove 1989). Philosophers today hold a wide range of opinions regarding the value of land apart from utility toward humans, but many philosophers (Leopold 1949; Naess 1973; Taylor 1986; Lovelock 2006; Sullivan 2013) make allowances for the protection of at least some land for scientific, aesthetic, or public purposes. Attitudes toward land use that incorporate environmental ethics can help to ensure a balance between commercial, scientific, political, and aesthetic applications of space resources.

4.2 Existing Treaties for the Use of Space

Space colonization, with Mars the first likely destination, challenges historical attitudes toward land use and highlights a need for a more sophisticated policy for the future. An unclaimed planet such as Mars still contains finite resources, and it is unclear from historical precedent how these resources should be allocated to the nations of Earth. A complete solution to this problem is lacking, but existing treaties that govern the use of space resources provide a place to begin developing a sustainable policy for the peaceful colonization of Mars.

A shift in attitude toward land use is reflected in present-day treaties governing the ownership of Antarctica. Although several nations have made territorial claims to parts of Antarctica, these claims were suspended under the Antarctic Treaty of 1961. The original treaty reserves Antarctica exclusively for peaceful purposes in order to promote international cooperation in scientific investigation in Antarctica, and the 1991 Antarctic Protocol extended this environmental protection to Antarctica by placing a 50-year moratorium on mineral exploration and mining activities (Rothwell 2000; Ehrenfreund et al. 2013). Antarctica is an example of how land use attitudes have changed, at least somewhat, in order to consider that land might provide public benefits that outweigh private interests. For at least the foreseeable future, Antarctica will remain a scientific oasis free from commercial interests or political control. The Antarctic Treaty System provides a model for Mars colonization that differs starkly from the philosophy of Locke and the Homestead Acts. A policy analogous to the Antarctic Treaty System would allow for the peaceful use of Mars for scientific purposes but likely would prohibit the establishment of permanent Martian colonies or programs to alter the Martian environment through terraforming. Such a policy may be favored among scientists and would distinguish Mars as a reserve for scientific research only.

The Outer Space Treaty of 1967 was drafted by the United Nations Committee on the Peaceful Uses of Outer Space (COPUOS) and has been signed and ratified by 102 nations including the US, UK, Russia, India, and China. This treaty explicitly forbids nations from claiming the right to occupy lands on the Moon or other planets, stating that, "Outer space, including the Moon and other celestial bodies, is not subject to national appropriation by claim of sovereignty, by means of use or occupation, or by any other means". The treaty establishes space as "the province of all mankind" that should be "free for exploration and use by all states without discrimination of any kind, on a basis of equality" with "free access to all areas of celestial bodies", clearly indicating that Mars and other celestial objects cannot be appropriated by a sovereign nation (Collins 2008; Ehrenfreund et al. 2013). This establishes a similar provision as with the Antarctic Treaty System that prohibits national claims to land; however, the language of the Outer Space Treaty is sufficiently vague that it does not explicitly discuss implications for individuals or multinational corporations who venture into space colonization.

The Moon Treaty of 1979 also was drafted by COPUOS, but this treaty was ratified by only 15 states, none of which are space-faring nations—the US, UK, China, Russia, and most of the EU have not signed the Moon Treaty. This treaty describes space resources as the "common heritage of mankind" and requires "equitable sharing" of the resources or derived benefits. The concept of public ownership reflected in this treaty parallels the language in the Law of the Sea convention that prohibits ownership of the world's oceans (Collins 2008; Ehrenfreund et al. 2013). Regarding Mars exploration, the treaty would permit samples of Martian rocks or minerals to be collected, which may provide an allowance for mining activities and mineral rights. However, the Moon Treaty considers any property that resides on the surface of Mars as a "fixture" that should remain open to common use by any exploring parties; a nation or corporation that invests in a Martian mining expedition can potentially make claims to the mineral rights of Mars but would have to allow others to use their infrastructure. Such a policy could act to deter investment in Mars exploration, but the Moon Treaty has not been ratified by any nations likely to colonize Mars in the near future.

Private plans for the colonization of Mars are being developed by several ambitious groups, but existing treaties are insufficient for governing these activities. If Mars colonization is the project of an individual nation, then the Outer Space

Treaty explicitly prohibits that nation from establishing permanent, enduring cities or otherwise claiming land for its citizens. However, individual or corporate colonization of Mars could follow the model of Locke and the Homestead Acts by funding private ventures that establish colonies independent of any particular nation. It is unclear whether or not the provisions of the Outer Space Treaty apply to this sort of private activity, and the lack of any further clarification could result in a scenario where private multinational corporations have more permanent access to Mars than government-operated space agencies.

If humans someday wish to colonize Mars, rather than keep it purely as a scientific reserve, then a new attitude toward land use is required (Brennan 1990; Almár 2002; Gruner 2005; Collins 2008; Haqq-Misra 2012; Ehrenfreund et al. 2013). The Outer Space Treaty mandates peaceful uses of space resources and prohibits national claims to sovereignty of these lands. Likewise, the Antarctic Treaty System provides a framework for the peaceful and scientific use of land, but it also prohibits ownership or control of any territories. It seems that only the philosophy of Locke and the tradition of the Homestead Acts allow for individuals to expand into undeveloped land and stake their claim. Yet this philosophy is problematic for Mars because it favors those with the economic means of financing space travel and because the Martian landscape is still finite. History has shown that even when resources or land appear to be abundant, humans can rapidly take advantage of these to the fullest. If human exploration of Mars does lead to permanent colonization, then careful planning for land use will be needed to help prevent the shortsightedness of the past.

4.3 Land Use Options for Mars

A robust policy for the use of martian land and resources needs to be established prior to the first arrival of a human on Mars. Such a policy must directly address the Outer Space Treaty, either by establishing Mars as a scientific reserve (following the Antarctic Treaty System) or by amending the Outer Space Treaty to include provisions that explicitly allow for colonization of Mars. Scientific research has guided Mars exploration thus far and likely will continue to be a strong force in developing martian colonies. Private corporations have also taken an interest in Mars exploration, which continues to generate discussion about the need to balance between scientific and commercial interests in space exploration (Ehrenfreund et al. 2013). The lack of explicit policies through enforceable international treaties will otherwise leave open the possibility of Mars colonization being dominated by wealthy individuals or multinational corporations that can operate beyond the confines of a single sovereign nation. Existing models, such as the Antarctic Treaty System or the Law of the Seas, provide examples that could extend to Mars, but the unique challenges of space exploration will likely require the consideration of new ideas for how to best allocate land resources.

One solution to this problem is to implement a "martian use tax" on any nations or corporations embarking on Mars exploration or colonization (Collins 2008). This policy would place martian resources under control of those for whom its value is greatest and would satisfy the "equitable sharing" clause of the Moon Treaty by distributing the benefits derived from martian exploration to all citizens of Earth. A portion of the use tax could be reserved for scientific and preservation efforts, and some areas of the planet even could be kept off-limits from colonists as nature reserves. Because some nations possess space-faring technology while others do not, the "martian use tax" policy provides a redistribution mechanism by which the benefits derived from space exploration are distributed equitably to the nations of the world.

Another solution to land use on Mars is to allow "bounded first possession by landfall" (Collins 2008), which allows a limited extent of martian land ownership. Under this policy colonists would be allowed a restricted form of first possession, where the claimed area is much smaller than the total martian surface. The first nation to reach Mars would still have their choice of location, but the radius of their claim would be of a relatively small extent (e.g. 100 km). This would leave opportunity for other colonists to make similar claims, perhaps limited to a few colonies per nation, while still keeping most of the martian landscape undisturbed. This policy might be difficult to enforce, for example, if individual colonies grow beyond their permitted radius, but the "bounded first possession by landfall" policy would provide a way for current Mars colonization efforts to proceed with minimal modifications to existing treaties.

An even more extreme option is to consider the possibility of liberating Mars (Haqq-Misra 2014). This would involve adopting the concept of planetary citizenship, so that colonists who permanently settle on Mars become citizens of the red planet who can no longer own property or hold any interests on Earth. This distinction between "Earthling" and "Martian" citizenship will allow colonies on Mars to develop into a second and independent experiment in human civilization. This option would satisfy the Outer Space Treaty by allowing Mars to benefit all of humankind while not favoring any sovereign nation, individual, or corporation. Allowing the citizens of Mars to dictate their own policies for land use will allow colonists to learn from the examples of human history while also addressing the unique concerns that will arise in the development and expansion of colonies on Mars. Deciding today to liberate Mars to its future human inhabitants—with all controlling interests of Earth superseded by the interests of the Martians—could provide a very real way to transform our understanding of land use and find a practical and sustainable way to share the resources of space.

The colonization of Mars will be a pivotal moment in the history of humanity, and the way we decide to approach the use of land and resources on Mars will set a precedent for space exploration beyond. We can choose to follow existing models for keeping Mars as a scientific and nature preserve, but such a decision would fundamentally limit the possible expansion of human colonies into space. Creative solutions that fit within the boundaries of the Outer Space Treaty could allow space colonization to proceed with minimal legal intervention, but not all such solutions will be satisfying to all stakeholders. History has shown the failing limitations of Locke's philosophy and the practice of homesteading, but it is unclear how to resolve the problem of land ownership and use in a finite martian landscape. Precisely because space exploration is an expensive endeavor, it is critical to not overlook the investment costs of the agencies that have enabled the exploration of space. Sustainably sharing the resources of space will require that we look beyond our current notions about the use and ownership of land to find practical solutions that can guide the first steps toward Mars colonization.

References

Almár, I. (2002). What could COSPAR do to protect the planetary and space environment? *Advances in Space Research, 30*, 1577–1581.

Brennan, W. J. (1990). Space colonization and the law. *Harvard Journal of Law and Technology, 3*, 7–12.

Callicott, J. B. (1989). *In defense of the land ethic: Essays in environmental philosophy*. New York: State University of New York Press.

Collins, D. (2008). Efficient allocation of real property rights on the planet Mars. *Boston University Journal of Science and Technology Law, 14*, 201–219.

Ehrenfreund, P., Race, M., & Labdon, D. (2013). Responsible space exploration and use: Balancing stakeholder interests. *New Space, 1*, 60–72.

Gruner, B. C. (2005). A new hope for international space law: Incorporating nineteenth century first possession principles into the 1967 space treaty for the colonization of outer space in the twenty-first century. *Seton Hall Law Review, 35*, 6.

Hardin, G. (1968). The tragedy of the commons. *Science, 162*, 1243–1248.

Hargrove, E. C. (1989). *Foundations of environmental ethics*. New Jersey: Prentice Hall.

Haqq-Misra, J. (2007). The power of our myth. *Astrobiology, 7*, 712–713.

Haqq-Misra, J. (2012). An ecological compass for planetary engineering. *Astrobiology, 12*, 985–997.

Haqq-Misra, J. (2014). The transformative value of liberating Mars. http://arxiv.org/pdf/1404.2315v1.pdf.

Leopold, A. (1949). *A sand county almanac, and sketches here and there*. Oxford: Oxford University Press.

Locke, J. (1690). *Second treatise of government*. Project Gutenberg, Salt Lake City, UT

Lovelock, J. (2006). *The revenge of Gaia: Earths climate in crisis and the fate of humanity*. Basic Books.

Naess, A. (1973). The shallow and the deep, long range ecology movement: A summary. *Inquiry, 16*, 95–100.

Rolston, H. (1988). *Environmental ethics: Duties to and values in the natural world*. Philadelphia: Temple University Press.

Rothwell, D. (2000). Polar environmental protection and international law: The 1991 Antarctic protocol. *European Journal of International Law, 11*, 591–614.

Sullivan, W. T. (2013). Planetocentric ethics: Principles for exploring a solar system that may contain extraterrestrial microbial life. In C. Impey, A.H. Spitz & W. Stoeger (Eds.), *Encountering life in the universe: Ethical foundations and social implications of astrobiology* (pp. 167–177). Arizona: University of Arizona Press.

Taylor, P. W. (1986). *Respect for nature: A theory of environmental ethics*. Princeton: Princeton University Press.

White, L. (1967). The historical roots of our ecological crisis. *Science, 155*, 1203–1207.

Chapter 5
The Price of Air

Adam H. Stevens

Abstract The amount of liberty available to colonists in extra-terrestrial settlements will be determined by a number of factors. At a fundamental level, the availability of oxygen will influence how the colony develops and how this affects the populace. In this essay, the economic factors that will decide oxygen availability are investigated. Firstly the context of colonization through history is described, focusing on how economic factors affected colonists. Then the three potential situations; where oxygen is either scarce, controlled by market forces, or abundant; are described, and the implications for the colonists of each situation investigated. Depending on the economic system that arises in the colony, individual colonists may be subject to centralized control from the state or corporations, and may have to alter their consumer behavior according to oxygen availability. Oxygen scarcity might also lead to social stratification in the populace, caused by differences in an individual's genetics or economic situation. Finally the growth of any such extra-terrestrial settlement will be controlled by the availability of oxygen and willing colonists, which will mean that colonization of other planetary bodies will not be able to proceed until technological development makes it economically viable.

Keywords Economics · Resource scarcity · Colonization · Life-support · Necessary labour

5.1 Introduction

There are many things that will decide how much liberty the eventual colonists of other planetary bodies will have access to, but the one thing that distinguishes these environments from society on Earth is that on the Moon or Mars even the very air needed to breathe will have a price. In this essay I will examine the factors and

---

A.H. Stevens (✉)
The University of Edinburgh, Edinburgh, UK
e-mail: adam.stevens@ed.ac.uk

C.S. Cockell (ed.), *Human Governance Beyond Earth*,
Space and Society, DOI 10.1007/978-3-319-18063-2_5

influences that will determine the price of this air, and therefore investigate what will determine the liberty of the colonists at a basic level.

To begin, I must make a number of assumptions about future colonies. For any 'outpost' constructed on another planet to truly make the transition into a 'colony' there must be some reason for it to do so. While there may be any number of reasons for colonists to decide to leave their home world, in order for a colony to be sustainable in the long-term it must be both self-sufficient (in that it cannot rely solely on imports of life-critical materials from Earth) and must be able to produce a surplus of some kind of commodity, which might be material or otherwise, above that which the colony itself requires for self-sufficiency and is of worth to someone outside of the colony itself.

A simple, but far from exhaustive, example would be to look at the Moon. Colonization of the Moon will only move forward if it is possible to produce enough oxygen and water there to sustain the lives of the colonists (food could, however, be imported) **and** if there is some reason to decide to live on the Moon, given that it is unlikely to be as hospitable as living on Earth. This reason could simply be a lack of living space on Earth, or something more economically driven such as the opportunity to generate income producing solar power, but without a reason to move there, the risks of life away from our home planet are too great to instigate settlement.

It is important before proceeding to define what is meant by 'price'. In this context, the price of a commodity is not necessarily the monetary cost, but could be an equivalent amount of barter goods or labour. The exact form of how trade will work in future colonies is unclear. The colonists of the future may choose to adopt a completely different way of approaching trade, but the concept of 'price' to an individual will still remain, no matter what system is in place.

The economic environment of an extra-terrestrial colony is difficult to predict. It will depend on many factors that are currently unknown, including advances in technology in the intervening time, the availability of resources and the ease of their extraction. To begin to understand what will shape the colonial economy, we might look at the history of colonization on the Earth.

5.2 Historical Context

One obvious analogue to extra-terrestrial colonies is the colonization of the North American continent, which has a long and complex history. Initial exploratory expeditions were followed by failed attempts at colonization as far back as Viking settlers or the Roanoke colony, which did not have sufficient backing to become self-sustaining. Later colonists, who had stronger economic and material backing and were able to utilize local resources, including the indigenous population who were so important for the Plymouth colonists, were able to make their colonies self-sufficient and, eventually, grow and expand them.

Of the initial thirteen North American colonies, some were primarily composed of people fleeing religious oppression in their homelands (Rhode Island, for example) whereas some were mainly concerned with exploiting natural resources for export to their home nations, most notably the British Empire. Each colony had its own local government, but taxation and trade policy was dictated by the British government, and inter-colonial trade was heavily restricted, with the majority of imports and exports having to come and go from other nations in the Empire through English ports. Before the American revolution, the 13 colonies were dependent on trade with the Empire, but as the colonies grew and became more self-sufficient and more able to produce goods of high economic worth this became less of an economic driver, and the revolution itself was probably a symptom of the turning point from the colonies being dependent on the Empire to being self-sufficient economic entities.

Thus the economic history of the American colonies has some analogues with the likely development of extra-terrestrial colonies. The North American continent was a hostile one. Quite apart from the indigenous population there were colonists from both France (in the north) and Spain (in the south) that were hostile to expansion. Yet there were abundant material resources and the ability for colonists to become self-sufficient and produce goods of worth that could be sold. The New England colonies were financed, commercially and by the state, but return on investment was absolutely expected, and production was heavily encouraged.

Pre-revolutionary development was rapid. Economic and population growth happened quickly in nearly every colony, increasing both economic and military security. One of the biggest issues that faced the colonists was interference from the colonial government. While rights were not severely restricted, economic restrictions meant that colonists, producing tobacco for example, made far less profit than they would if able to sell on the open market due to import and export fees and the necessity of intermediary shipping through English ports. This, as well as a number of taxes that were imposed by the government in order to offset what they saw as the investment in the colonies, was a major influence in starting the American Revolution.

Return on investment meant that there was both an incentive for the state and commerce to encourage colonists. More colonists meant that more land could be occupied, which prevented expansion by other colonizing nations, provided a larger labour force and increased production. People sailing across the Atlantic often did so on contracts that would see them indentured to their sponsor for a given term, after which they were free to seek other employment (Atack and Passell 1994). Often these contacts were not overly exploitative and therefore both parties came out ahead, with the sponsor gaining cheap labour and the colonist gaining a journey to the new world that they would otherwise have unable to finance.

Post-revolutionary US expansion also holds similarities with colonization of other planets. While early settlement was often due to adventurous individuals 'homesteading' land further to the west, true colonization only happened following state supported infrastructure development, with railways being the most important. This development allowed individuals to move west in an attempt to 'make their fortune', exploiting the uncertain availability of highly sought after resources, and settlements regularly grew up around this practice, helped by the sheer availability of land in these areas. Because of the strong central government assembled after the revolution, the ability to leverage trade with other nations was increased over that of the individual states doing so, and resulted in increased revenues, boosting the economy nationally.

The colonization of the American continent by European nations was essentially driven by the same reason as those colonies on many other continents—to increase or diversify trade, especially the availability of luxury goods. For example, the northern American colonies provided a large amount of furs and pelts to the European markets, and tobacco, sugar, rum, and spices were commonly produced in colonial nations.

The loss of the American colonies in the War of Independence caused Britain to look for other places to produce similar goods, and the colonization of Australia was a direct result of this. In fact, the American colonies had been supplemented by the transport of convicts far before the Australian colonies, and the extreme geographical disconnect between Britain and its colonies meant that such transport was genuinely frightening for those receiving it as a punishment: Australia "might as well have been another planet" to many of the population (Hill 2008). This psychological disconnect due to geographical distance can only become more pronounced when the colonists are living on a different planet, and the true impact of leaving Earth on the colonists will be unknown until it happens.

Judicial incitement of migration also has a rich background in the eastward expansion of Russian colonists into Siberia. While early 'settlers' were often sent east as punishment for crimes, the culture eventually changed so that the majority of travellers were fleeing potential punishment, such as debtors fleeing their creditors (Lemon and Pollock 1980). The idea of 'penal colonisation' raises an interesting prospect for future extra-terrestrial colonization. While it may seem unsavory, cheap labour from criminal elements of society might be worth considering for the early phases of extra-terrestrial settlements, at least while living conditions and security of essential resources remained poor during the early stages of colonization. This might echo Heinlein's *The Moon is a Harsh Mistress* a little too accurately for some tastes, but might become a necessity depending on the economic situation in the future.

The question of whether extra-terrestrial colonies will offer the same economic incentives as throughout history is an important one, and it may be that we see a more general change in the mechanisms of colonization when humans begin to live beyond the Earth. No terrestrial colony was initiated due to lack of carrying capacity in the home nation (except perhaps to ease the burden in jails), but this may not be true in the future. Lack of viable living space and increasing resource scarcity on Earth may drive the colonization of other worlds, significantly changing the economic context of these colonies and making it more likely that future colonists will "put up with" or accept more difficult economic conditions, given that they have no choice otherwise.

The comparisons we can draw between historical terrestrial colonization and future extra-terrestrial colonization are potentially useful, but ultimately shallow. While in many cases the environments that previous colonists faced were hostile, none offered the direct threat to life that living on the Moon or Mars will in the future.

The nearest we can get on Earth would be to consider the settlement of extremely hostile regions such as Antarctica. Yet this comparison has little to offer us in terms of economic development. Human presence in the Antarctic, despite more than a hundred years of exploration, has not moved beyond living in what must be classed as scientific outposts, and it is possible that true colonization of the continent will never happen, depending on whether the natural resources of Antarctica become valuable enough.

Colonization of the Moon or other planetary bodies will be fundamentally different to any other colonization in our history and as such it may be difficult to use historical precedent to predict or inform the social, political or economic development of these colonies. The most extreme difference, the fact that the environment of these colonies will be inimically hostile to life, and the lack of immediately breathable air, has already been identified as a major influence on the way extra-terrestrial colonization will progress (Cockell 2013). In contrast to historical colonies, population growth will reduce the material security of the extra-terrestrial colony and serve to increase the pressure on life support systems, meaning that rapid growth will be highly undesirable for extra-terrestrial colonies.

The questions that we must ask then, as we imagine our future colony, are manifold. Who will the colonists be? Will they be explorers, or radicals, or perhaps the persecuted, or the criminal? In each of these cases the reason for their dislocation from their home planet will be different and create a different context for colonial development. What will the daily lives of the colonists be like? Will they be easy, or will they be hard and filled with constant labour, or with sufficient time to spend on other pursuits? What exactly will the labour of the colonists be used to achieve or produce?

Really though, the major underlying question that I seek to investigate here is simple: what will be the price of the air that the colonists breathe? The answer to this question is what will determine, at the most basic level, the amount of liberty available to the individual colonists. There are three potential over-arching answers to this question, with multitudes of other factors that will determine the exact solution. If air is incredibly hard to come by, requiring every waking hour to produce or procure enough to stay alive, the situation will be very different to if air is easy to make, or abundant enough that no single person has to worry about their supply. Between these two extremes will exist a greyer area, where the price of air might fluctuate or be dependent on who exactly is "buying" it. I will attempt to examine the three situations of oxygen scarcity, market controlled availability and oxygen abundance in detail.

5.3 Scarcity

For a colony to be economically viable, it must not depend on the Earth for its air, implying that in-situ resource utilization will have to be exploited. The availability of important resources, including oxygen, is highly variable across the bodies of our solar system, being freely on some but locked away in different forms on others. Taking the example of the Moon, it is impossible to say with our current knowledge how much easily accessible oxygen exists on or near the lunar surface. If water is not abundant (perhaps only available in permanently shadowed craters) then the major source of oxygen might be the rocks of the lunar regolith. If so, then it is further impossible to say how 'easy' it will be to extract the oxygen from these rocks, at least with current technologies.

Therefore, it is possible that it could take an extreme amount of effort and physical labour to produce oxygen on the Moon. The production process might only be available to corporations with sufficient capital to develop and produce the necessary machinery, or it might be possible for an individual to build and maintain their own "air plant". If this air plant requires constant attention and labour from its owner (such as if it were based on photosynthetic plants, for example), then life on the Moon for these people would involve nothing more than waking up in the morning to produce enough air to keep them alive for another day, and then repeating the cycle. While there are many places on the Earth where people must survive by spending their time growing just enough food to sustain themselves, this situation is not quite the same as the potential drudgery and lack of security that a lunar colonist could be subject to. Why would someone choose to leave the Earth to fly away only in order to live a life of endless insecurity, not knowing if tomorrow will be their last day? Only for those that had no choice would this be an option.

An important point to realize in this case is that oxygen is a renewable resource. If the carbon dioxide excreted by colonists can be captured, then labour, resources or energy can be expended to recycle it back to usable oxygen using fairly simple processes. Once a colony has reached the critical point of producing enough oxygen to self-sustain and can recycle its oxygen sufficiently efficiently, it should be relatively easy to maintain reserves above the critical point, and additional production would therefore go towards moving the colony beyond the situation where oxygen is scarce.

If the air production machinery were corporate, but still marginal, then a situation may arise similar to that in colonial America. People might choose to live on the Moon on the promise that they can "work off" their debt of air to the corporation. But if this is possible, then more colonists will be required once these "debts" have been worked off, and if the production of oxygen were marginal enough, then this population growth would quickly create a potentially catastrophic situation. If it is not possible to production more than the "bare minimum" of oxygen for the population, then what incentive does the corporation have for developing and running oxygen producing technology, if their only customers are their own workers? Such an industrial situation would be completely unsustainable. It might exist in an early phase of colonization, if the levels of available resources were unknown, but would quickly reach a dangerous point unless the efficiency of extraction could be increased, or more resources found.

In the extreme example of an absolute oxygen scarcity, there would be no surplus labour, only necessary labour, which would be just as economically unviable as if there were no oxygen available in the first place. Colonists would be required to live 'marginal' lives with little security, long work hours and little time to exploit what little liberty they might have. The price of breathing to the colonist in this situation is absolute, and they have no additional worth to offer. This situation represents one potential extreme of extra-terrestrial colonization, and a very unattractive one, but one that is unlikely to occur, as if it did colonization would not continue.

5.4 Market

A less extreme and more realistic scenario is the case where production of breathable oxygen does not use up all surplus labour. Perhaps colonists have to devote a few hours a day to go to work in the oxygen 'factories' or maintaining their own air miners. Increased availability of oxygen would allow stockpiles to be built up, which would be critical for the expansion of the colony, depending on the efficiency of any recycling system. In reality, it is unlikely that extraction or production of oxygen will continue at exactly the same rate over long timescales. The availability of specific resources or any number of other factors will change over time, potentially leading to 'lean periods' that require these stockpiles to be used, slowing growth. The specific values of extraction and production rates and how they change over time will determine the true amount that can be stockpiled, which in turn determines the exact amount of manageable growth that can be sustained, and how much of an emergency buffer can be sustained to prevent danger to life during lean periods.

In the extra-terrestrial context, oxygen can be considered as completely inelastic commodity—every single colonist would need to purchase the same amount of oxygen for themselves in a given period (accounting for changes in their own activity or the fact that it might be free in public places, for example), and therefore demand does not change with time, if the population stays constant. Yet the supply (and therefore the price) of oxygen will fluctuate, at a rate determined by the exact situation of production in the colony, forcing colonists to be in what might be a very uncomfortable situation of not knowing how much they will need to spend on oxygen next month, a more extreme version of what many people with low disposable incomes are subject to across the Earth already.

However, the argument that oxygen is a perfectly inelastic product is subject to some caveats even considering potential fluctuations, and this will lead to some novel forms of social stratification. Suppose that a colonist is down on their luck and cannot afford to purchase their required air and water for the month. They could choose to purchase less their needed oxygen and ration themselves to breathe lower than the standard 21% oxygen mix. Humans can operate over long periods of with oxygen mixing ratios down to around 16 % at atmospheric pressure (Alling et al. 1993), meaning that it would be more than possible for someone to reduce their "O_2 spend" over a given period. In fact, after a reasonably short adaptation period, people breathing lower than 21% oxygen suffer no obvious lethargy or deficiency in capability, meaning that some people could choose to do this long term in order to save their wages. People moving in and out of a low oxygen 'diet' as demand varies would, however, suffer through more of these adaptation cycles.

But social stratification would go even deeper than simply those looking to reduce what they have to spend on the bare essentials. An individual's aerobic fitness affects how much oxygen they require to stay alive. People with increased fitness wouldn't need to spend as much on air, and even if the difference were only minor, this would begin to stratify the colonial population economically based on their fitness. While this would, of course, encourage people to exercise more in order to increase their aerobic fitness and reduce their spend, genetic differences and other factors that also play into aerobic fitness could create a complex new social class based on nothing more than how much you breathe. A worrying result of this stratification could be the rather dystopian idea of 'Oxygen Bars' where those forced to breathe lower percentages of O_2 could spend a little extra money to feel the elation generated from breathing a richer oxygen mix. Or, in a similarly dystopian way, a nefarious state police apparatus or corporations could control the percentage of oxygen in the generally available air in order to control the populace using targeted periods of reduced oxygen adaptation.

In a colony where air is neither a completely scare or a completely abundant resource, the availability of breathable oxygen, and therefore its price to the consumer, will generate social stratification based on very real physiological issues, and allow those able to stockpile resources to create a potentially life-threatening situation for colonists with lower disposable incomes. One way to avoid this would be for the state to ensure (either by legislation or nationalization) that the supply of air was separated from market influences as far as possible. However, even complete nationalization would not be able to completely disconnect the price of air from the market, meaning that even in a situation where the air supply is not commercially controlled the fact that oxygen becomes a commodity will generate what might be an unpleasant social structure.

5.5 Abundance

The other extreme situation that may arise might be called "supra-scarcity," where oxygen is freely and easily obtainable. Perhaps technological breakthroughs enable everyone to have a personal oxygen-producing unit that provides enough to keep them alive while requiring very little maintenance or input from the user. In this case it would be possible to stockpile large amounts of oxygen (whether by the individual, in groups, by corporations, or by the state), increasing the security of the air supply, but also importantly increasing the security of the individual's liberty. With an abundant supply of oxygen, people would be free to use their time without worrying about whether they will be able to breathe tomorrow. With large stockpiles of oxygen it would be difficult for any one entity (whether state or commercial) to monopolise the air supply and therefore harder for any such entity to impose controls on the populace, whether economically or militarily.

In theory this last situation should bring the political and economic environment of the extra-terrestrial colony more closely in line with that of 'normal' life on Earth. Yet it is difficult to imagine that any real extra-terrestrial colony will have an abundance of oxygen available to the colonists, unless we invoke more futuristic and global terraforming processes. In reality, the specific environment itself will decide the exact circumstances of oxygen availability. In a lunar colony, for example, it is more likely that oxygen will be drastically scarce and therefore that whatever entity controls production of oxygen will have major economic and political power, to the extent that the liberty of the individual colonist will be severely limited. On Mars, though, where oxygen could be relatively easily extracted from the carbon dioxide atmosphere, more complex market forces, rather than a state or commercial monopoly, will most likely control the availability of oxygen.

In both realistic cases an important and interesting problem arises. If oxygen is not freely available, has an associated 'price' that might be quite high, and perhaps is not adequate to support a growing population, then there will always be some people that could be seen as not 'contributing' their fair share to the sustenance of the colony. To take a simple example, artists, who would certainly enrich any extraterrestrial society, and may in fact be critical to its survival (Beattie 2014), could be seen as a 'drain' on the life support resources of the colony and marginalized because of this. Non-productive colonists might form another class of society looked down on others even more than they are on Earth, an extreme version of the view that people such as these are a drain on the resources of society, but a very real and important one. To promote rich and vibrant society in these situations that could sustain people that did not directly contribute to the life-support apparatus would therefore require some kind of 'societal surplus' of oxygen, further placing constraints on the population of the colony and most likely requiring severe planning and forethought.

5.6 Summary

In all of situations described above, oxygen turns from a freely available resource (as on Earth) to one of high value, more akin to mineral or hydrocarbon resources on Earth, but with one major difference—every single member of the populace is dependent on it. This shift is the fundamental difference between colonization of dangerous parts of the Earth and the colonization of extra-terrestrial bodies. This important difference is key to defining at a basic level the liberty available to future colonists and that the specific situation that will arise in these future colonies will depend on a number of currently unknown factors, most importantly the availability of oxygen in a given colony on a given planetary body, but also the level of technological development at the time of colonization. While these situations have some analogues with terrestrial colonization throughout history, they perhaps bear more resemblance to places on Earth where other critical resources are not abundant. The social and political structures that arise in such colonies may be more similar to those of places where food or water is scarce. Such adaptations can include community level developments such as unregulated private or black markets where state systems have failed (Mohieldeen 1999), or personal level strategies such as minimizing usage, but human societies have shown remarkable resilience in adapting to environmentally-driven resource scarcity.

The potential events described above would most likely form the 'pioneer' phase of an extra-terrestrial colony. For the colony to move beyond this phase into a more stable one, it would have to be in the market or abundance situation. If it were subject to oxygen scarcity, then measures would have to be put in place to increase the availability of oxygen before the colony could grow. This could either be by increasing the productive capability of the colony, perhaps by utilizing technological advancements. Whatever the desire to move the colony into a self-sustainable phase, there would have to be a fine balance between growth and production. If it were to grow too fast, a colony would quickly increase the danger to its citizens, and any such growth would have to be carefully managed in concert with improvements in infrastructure and production capability.

It is possible that technological developments could make the majority of the issues described here partially irrelevant, but the process of these developments coming about and their eventual application in an extra-terrestrial context will still require investment from either the state or corporations and therefore some of the issues will still apply. Technology has been used many times on Earth to reduce the pressure from resource scarcity, a notable example being the Green Revolution caused by the development of high-yielding varieties of grains and other technology, which increased food security for millions of people.

The shift in economics caused by the criticality of personal oxygen supplies will ultimately ask an important question of any future extra-terrestrial colonists—why do they want to leave Earth? If for fear of oppression, fleeing justice, or as an alternative to imprisonment, then colonists may be more willing to cope with the harsh realities of a life where your air supply is not guaranteed. We can compare these pioneers with the settlers of the North American continent, or the Australian mainland, who had to undergo significant hardships, but were ultimately rewarded with more liberty than they had before their journeys. On the other hand, the harsh realities of life on the Moon or Mars will be much harder to sell to those pioneers seeking their fortunes beyond the Earth, at least until we have a much better idea of the riches that they might be able to exploit. Finally, if extra-terrestrial colonization proceeds out of nothing more than necessity, perhaps because the Earth runs out of carrying capacity, then these harsh realities would become simple normality for the colonists.

This essay aims only to address the most basic of points in extraterrestrial colonization, that of whether colonists will be able to afford to breathe. It has not touched on many important technological, political, economic or sociological points, including the dangers of storing reserves of oxygen, or the possibility of inter-colony oxygen trade, but it has shown that planning for oxygen management in extraterrestrial colonies will be critical and that any systems involved, and the growth of the colonies, must be carefully managed, at least while these colonies are developing to a stage where oxygen is abundant.

However, this discussion does raise the depressing thought that, as of our current understanding of potentially colonisable bodies in the solar system, there is no sufficient economic argument for human colonization. Until significant resources on other planets that are rare or becoming rarer on the Earth are discovered, extraterrestrial colonization will be confined to scientific or 'prospectors' in commercial outposts. There will not truly be colonies until there exists a reason for people to want to live there despite the associated dangers.

References

Alling, A., Nelson, M., & Silverstone, S. (1993). *Life under glass: The inside story of Biosphere 2*. Oracle: Biosphere Press.

Atack, J., & Passell, P. (1994). *A new economic view of American history: From colonial times to 1940*. New York: Norton.

Beattie, A. (2014). Extraterrestrial liberty and creative practice: A firsthand experience of an imagined future. In C. S. Cockell (Ed.), *The meaning of liberty beyond earth*. Cham: Springer International Publishing.

Cockell, C. (2013). *Extraterrestrial liberty: An enquiry into the nature and causes of tyrannical government beyond the earth*. Ferris: Shoving Leopard.

Hill, D. (2008). *The brutal truth of the first fleet*. Australia: William Heinemann.

Lemon, A., & Pollock, N. C. (1980). *Studies in overseas settlement and population*. New York: Longman.

Mohieldeen, Y. (1999). Responses to water scarcity: Social adaptive capacity and the role of environmental information. A case study from Taiz, Yemen.

Chapter 6
The Birth of a New Republic: Depictions of the Governance of a Free Moon in Science Fiction

Stephen Baxter

Abstract This essay, intended as input to the drafting of the 'constitution for an independent lunar base' included in the present volume, is a survey of the portrayal of the governance of independent human colonies on the moon in science fiction. Self-governance on the moon will have some challenges in common with settlements on other solar system bodies such as Mars, for example a supremely harsh natural environment and a reliance on essential communal systems. But the moon will present some unique challenges of its own, such as the economic and military domination of a nearby Earth, and the fact that it is likely to be the site of the first move for independence from Earth. In terms of fictional lunar polities, the classic work is Heinlein's *The Moon is a Harsh Mistress* (1966). Other writers have explored challenges including military, technological and economic.

Keywords Moon · Lunar colonisation · Governance · Extraterrestrial liberty · Science fiction

6.1 Introduction

Liberty on the moon has been explored in stories dating back to the origins of the modern science fiction (SF) genre. 'The Birth of a New Republic' by Breuer and Williamson (1931) is a conscious re-run of the events of the American Revolution set on the moon. On a crudely terraformed moon with two million inhabitants (p. 14), the war is fought out with huge atomic spacecraft and gaudy weapons, and in gigantic caverns inhabited by hostile 'Selenites'—who, in the story, parallel the role of the native Americans. Perhaps the most interesting political variation from the historical precedent, and with modern resonances, is that the Earth of the 24th century is dominated by a handful of huge corporations, not governments.

S. Baxter (✉)
British Interplanetary Society, London, UK
e-mail: sbaxter100@aol.com

C.S. Cockell (ed.), *Human Governance Beyond Earth*,
Space and Society, DOI 10.1007/978-3-319-18063-2_6

The lunar colonists seek independence, not from an imperial power like Britain, but from economic control by the 'Metals Corporation' (p. 13); in fact they strive to establish their own independent corporation rather than a new nation. Perhaps the historical parallels are too knowing, but the sweep of the piece is irresistible: 'I was soon within the glass walls of the city; as a young officer, I took part in the innumerable balls and banquets given in honour of the victors. But my heart was seldom with them. I thought only of the dark-eyed girl who was waiting for me in the little city far across the frozen lunar wastes' (p. 78). This was the first novel by Williamson in a career that would span decades.

Human societies and their governance systems do not develop in abstraction but in response to frameworks of geography, history and politics. For example Crawford (2014) has eloquently described how the US federal system of governance was an ingenious response to the situation in which the rebellious colonists found themselves, as they debated their new constitution in their handful of colonies on the eastern edge of a continent; it was a system able to expand to incorporate states yet to be founded in lands yet to be explored.

Independence-minded lunar colonists must find a means of self-governance in the context of a supremely harsh natural environment and the economic and military domination of a nearby Earth—a background without precedent in our history to date. Who established and funded these colonies? How many colonists will there be? What will they be doing on the moon? How will they survive? How will they be making their living? What relations will they have with human communities on Earth and elsewhere? What will their aspirations be, and what resources will they have to achieve those aspirations? The answers to such questions will frame the making of a lunar constitution, just as the circumstances of America in the 1780s framed the making of the US constitution.

Of course the lunar future is unknowable. But in this area as in many others the best relevant science fiction works, the products of trained and disciplined imaginations, provide a bank of thought experiments concerning the problem in hand.

6.2 Science Fiction and the Moon

General surveys of the science fiction field include the classic work by Aldiss (1986) and a more recent history by Roberts (2006). The online SF Encyclopaedia (Clute et al. 2014) is a fine, free and up to date resource.

Tales of the human colonisation of the moon are only a subset of a wider literature concerning the moon dating back millennia, and ultimately fading into preliterate folklore (for surveys see Spudis (1997) and the relevant entries in Clute (2013)). This literature reflects the increasing scientific understanding of the moon and technological progress concerning how it might be reached.

To early observers—especially pre-telescopic—it seemed possible that the moon, an island in the sky, might be a destination for travel, with air, oceans, and life. The Greeks understood that the moon is a spherical world at least to some extent like the Earth. Plutarch, of the first century AD, even believed that the moon was inhabited by people. This was reflected in the fiction of Lucian. Serious interest in the moon as a world in its own right, and perhaps as an abode of life, began in the telescopic era. In the seventeenth century Johannes Kepler speculated in fiction on how the heavens would appear from the moon, and on an ecology on the moon. In early works lunar societies were often used for satirical purposes, as dark mirrors of human societies.

By the time lunar voyages began to become a matter of engineering rather than fantasy, the moon had already been extensively explored telescopically, and the prospect of life, at least on the surface of the moon, was beyond credibility. Jules Verne's circumlunar travellers in *Around The Moon* (1870) observed only traces of life in deep craters on the far side of the satellite. H.G. Wells got around the deadness of the surface by imagining a lunar society *within* the body of the moon, in *The First Men in the Moon* (1901). Later pulp-era writers like Burroughs (1926), and indeed Breuer and Williamson (1931), would invest the moon with life, sometimes exotic and extravagant, often hidden beneath the surface or on the far side. But other contemporary writers used the moon as a symbol of desolation, disappointment and loss. For example in George Griffith's *A Honeymoon in Space* (1900) romancing interplanetary travellers find the ruins of long-dead lunar civilisations.

In the twentieth century, as the engineering of space travel was gradually developed, increasingly plausible fictional depictions of a first lunar voyage appeared, such as Clarke's *Prelude to Space* (1953), as well as quasi-realistic stories of lunar colonisation, notably Heinlein's 'Future History' tales (1939 onwards) of which more in the next section.

More fantastical possibilities for the moon have continued to be depicted (see Baxter (2001) (Chap. 7) for a survey), such as its being inhabited by aliens (Lewis 1945; Baxter 1998). Or, if the moon could not harbour its own life, perhaps it has been visited, as in Charles Chilton's BBC radio serial *Journey Into Space* (novelised from 1954) and in the adventures of Perry Rhodan (Scheer and Ernsting 1961). In the BBC's *Doctor Who* the moon has been used as a military base by the Daleks ('Victory of the Daleks' 2010), the Cybermen ('The Invasion' 1968) and others. Apollo astronaut Buzz Aldrin co-wrote a saga of alien contact in which an elaborate 'encyclopaedia' was left behind on the moon (Aldrin and Barnes 1996). The most famous lunar visitors, of course, were Arthur C. Clarke's Firstborn, who, in the film and book *2001: A Space Odyssey* (1968), buried a monolith millions of years ago beneath the Tycho dust.

Some writers have depicted the terraforming of the moon in the further future (Gribbin and Chown 1988; Baxter 1998). One of the strangest, and most beautiful, fates for the moon came in Brian Aldiss's *Hothouse* (1962). In the far future, giant spider-like creatures called traversers have learned to cross between Earth and moon: 'Back and forth the traversers could shuttle at will, vegetable astronauts huge and insensible, with Earth and Luna both enmeshed in their indifferent net. With surprising suitability, the old age of the Earth was snared about with cobwebs' (p. 17).

6.3 Living on the Moon

How and why will humans be living and working on the moon? SF has offered evolving answers.

At the time of their foundation lunar colonies will inevitably be reflections of terrestrial politics and economics. For example, the background to Clarke's *2001: A Space Odyssey* (1968) is an Earth riven by hunger and nuclear tensions (p. 37), and so the moon appears to be divided into nationally controlled 'sectors' (p. 49) with, evidently, many bases. The Soviets have a radio telescope observatory on the farside (p. 48). In the US sector, Clavius Base, built by the US Astronautical Engineering Corps (pp. 62–63), is a government operation, with 1700 staff of 'highly trained scientists and technicians' (p. 67). When Dr Heywood Floyd comes to the moon to investigate the strange alien monolith that has been discovered in Tycho, he brings a message from the President, and the Americans' momentous discovery is kept a national secret. However, despite the military and technical specialism, and although it has the feel of a 'hardened missile site' (p. 60), Clavius Base also looks to the future. It is intended to be the 'first permanent bridgehead' on the moon. All the 'necessities of life' are being extracted from the lunar rock; in a closed ecosystem the air is replenished in a 'hothouse' full of gen-enged plants, and food is produced with algae cultures: 'Clavius Base could, in an emergency, be entirely self-supporting' (p. 58). And the 'first generation of Spaceborn' is already being born here; Floyd reflects that 'the time was fast approaching when Earth, like all mothers, must say farewell to her children' (p. 64).

In many ways Clarke's Clavius Base anticipates many functions of a lunar colony, as explored both in fiction and in reality.

The notion of using the moon as a military base is an old one in fiction. In Heinlein's juvenile *Rocket Ship Galileo* (1947) there are Nazis on the moon, seeking to use the satellite as a military high ground to establish a new Reich. This idea was revisited to comic effect in the spoof movie *Iron Sky* (2012) which features a huge swastika-shaped Nazi base on the moon, preparing for a 'meteorblitzkrieg', an invasion of the Earth. In Murray Leinster's *City on the Moon* (1957), 'the [lunar] military missile bases … were the first foothold of mankind on any natural celestial body' (p. 6). This time the missile bases are American, and a kind of atomic peace has been forced on a resentful Earth. The plot is a lurid thriller about sabotage and murder, driven by Cold War tensions, but the detail of life on the moon is interesting.

In reality the early 1950s US space programme studies led by Wernher von Braun, and famously outlined in articles in the *Collier's* magazine (Ryan 1952) were heavily militaristic in tone. Von Braun wrote of the possibility of an armed space station, for example: 'In the hands of a peace-loving nation … the Space Station would be a deterrent which might cause a successful outlawing of war (p. 56) … And if we don't [build the Station], another nation—possibly less peace-minded—will' (p. 19). Cold War Studies of the militarisation of the moon soon followed.

Allen Steele's alternate-history novel *The Tranquillity Alternative* (1996) dramatises this early vision. The historical turning point comes in 1944, when Nazi Germany threatens America with a Sanger spaceplane (a development Steele explored at length later (2014)), and US aerospace development is accelerated, in Steele's story following precisely the timetable outlined in von Braun's *Collier's* articles. A US Space Force is established, and 'Atlas-C ferries' to orbit, based on von Braun's three-stage winged design, are flying by 1956 (p. 29). But even by 1968 the military logic is fading, and President Robert Kennedy campaigns to have the USSF give way to a civilian NASA organisation (p. 73). And there is general disgust at the exposure, in 1977, of a secret lunar base called 'Teal Falcon', set up in 1969 and containing six nuclear-tipped Minuteman missiles (p. 128). Steele notes in his afterword (p. 304) that he drew on USAF studies of second-strike nuclear deterrence based on lunar facilities dating back to the 1950s. The book is a fable of American political short-sightedness, as Steele sees it, concerning the possibilities of the space frontier.

Plausible uses of the moon came into much sharper focus after the Apollo missions. Not only had humans walked there but the post-Apollo science results gave a much clearer picture of the resources of the moon, which might include polar water, iron, magnesium, silicon and perhaps high-value minerals such as platinum, all of which had potential in supporting humanity's move off the Earth. In studies by such as O'Neill (1976), near-future moon colonies typically fed off water ice mined from the lunar poles, and extracted resources to supply such near-space operations as power satellite construction. For example the 'Stanford Torus' design (Johnson and Holbrow 1977), based on a 1975 workshop involving NASA Ames and Stanford University, was to be a space habitat in the form of a torus 1790 m across housing 10,000 people. The total mass would be $\sim$10 million tons, of which 9.9 million tons would be a radiation shield of lunar slag. The endeavour would require $\sim$150 people working in facilities on the moon. Helium-3, too, a valuable fusion fuel sparsely implanted by solar radiation in the lunar regolith, came to be seen as a worthwhile commercial target (Slyuta et al. 2007).

Fiction writers have used these visions of a new kind of Earth-moon economy to spin new stories of human life on the moon. Allen Steele's *Lunar Descent* (1991) is a story of blue-collar workers in a lunar colony: 'Living on the moon sucks, I gotta tell you' (p. 18). In the year 2024 (p. 6) Descartes Station is home to a maximum of 112 staff (p. 18). Living in huddles of 'mooncrete' buildings, the workers, on fixed-term contracts, labour to extract lunar resources for building powersats. A crisis comes when the water ice at the lunar north pole (p. 238) begins to run out. An unscrupulous management, already running down the station, tries to sell it off to off-stage Japanese. The situation is resolved when the 'moondogs' themselves arrange a buy-out of the colony. This is a very American fantasy, of downtrodden but heroic workers winning a new frontier through individual initiative and hard graft.

Ben Bova's novels of a near-future lunar colony, beginning with *Moonrise* (1996), and set against the wider background of Bova's saga of the colonisation of the solar system, tell a similar story to Steele's, of a lunar colony, Moonbase, set up for corporate purposes struggling for financial survival, and striving for autonomy. Nanotechnology, however, appears to transform the prospects for the base. With the new technology the colony could grow economically and become self-sufficient (p. 475). But on an inward-looking, religiously conservative Earth, nanotechnology is being banned (p. 463). Moonbase becomes the fulcrum of the struggle between conservatism and vision, between future and past. The story itself is *Dallas* on the moon: a saga of manoeuvring corporations, feuding families, intrigue, murder, heroic blue-collar workers and visionary entrepreneurs.

The scientific value of the moon, as opposed to its mineral riches, inspired Bova's *Farside* (2013). Set some years on from Bova's earlier lunar-saga books and with the independent lunar nation of 'Selene' prospering, this is a drama centred on an ambitious scientific project. In the radio shadow of the lunar farside a large radio telescope array has already been established: 'Farside … was the quietest place in the solar system for sensitive radio searches for intelligent life' (p. 17). But now a major optical telescope is to be installed to inspect an Earthlike planet in the Sirius system: 'The vacuum at the lunar surface was actually a thousand times thinner than the vacuum in Earth orbit. Lunar materials could be used to build the telescopes, and the moon offered a firm platform for them' (p. 18). (Indeed there are studies on the use of advanced spaceborne optical telescopes for studying exoplanets; see for example Tcherniavski (2014).) The plot concerns sabotage and murder driven by academic rivalry.

Often the moon is used as a prison, such as in the *Doctor Who* serial 'Frontier in Space' (1973). The lunar cities in Heinlein's *The Moon is a Harsh Mistress* (1966) were penal colonies. Robinson's 'The Lunatics' (1988) is a more surreal saga of a lunar penal colony, with memory-wiped convicts endlessly tunnelling within the moon in search of minerals.

The BBC's *Doctor Who* has featured some more imaginative uses of the moon. 'The Moonbase' (1967), written by the series' science adviser Kit Pedler, was purposefully set on the moon to resonate with the then-current interest in the Apollo programme (Howe and Walker 2003, p. 164). In the year 2070 'Moonport' is a domed base, manned by a small number (15) of multidisciplinary, multinational specialists. The purpose of the base is to control Earth's weather using a 'Gravitron', a gravity manipulator—an early example of what we might now call geoengineering. Unfortunately the Cybermen attempt to take over the base, and use the Gravitron to disrupt the Earth. In 'The Seeds of Death' (1969), set in the late 21st century, the moon is a hub of a global 'T-Mat' system—'Travel-Mat'—a teleportation-based system that is used for the transport of goods and people, under UN control. The moon serves as a natural communications satellite for the system. This time the Doctor has to thwart an elaborate plan by Ice Warriors from a cold and dying Mars to use the T-Mat to support an invasion of Earth.

A moon studded with many colonies, perhaps of different national or corporate origins and with diverse purposes, is not uncommon in fiction. In Gerry Anderson's TV series *UFO* (1969) the Earth is under attack by humanoid aliens. A Moonbase is the outer perimeter of a deep defensive network, with deep capabilities and a first line of defence in spacegoing interceptor craft. But we glimpse other bases, private and national.

Note however that bases on the moon may not hold large human populations. Steele's Descartes Station (1991) held only 112 workers. In the movie *Moon* (2009) a helium-3 extraction operation requires just *one* human operator; the rest of the operation, though extensive, is automated. The story hinges on the operator discovering that he is one of a series of short-lived clones; his 'shift' will end only in his death!

The discussion here so far has been utilitarian: the moon seen as a mine of resources. Could humans ever learn to love the moon—love it enough to fight for its independence? Of all the science fiction authors working in the period between the end of World War II, when the human race realised that for the first time it might have the practical means to reach space, and the Apollo landings when the moon was reached, nobody did more to delineate, mythologize and promote an expansive human future in space than Robert A. Heinlein. And an inhabited moon became central to Heinlein's celebrated 'future history' project, a vision of mankind's progressive development and spread into space.

Set in the late twentieth century, Heinlein's short novel 'The Man Who Sold The Moon' (1949) and its sequel 'Requiem' (1939) (the latter actually written first) presents the story of Delos D. Harriman, genius entrepreneur, who wheels and deals at the limits of his personal financial means and of the law, in order to fund the first trip to the moon, and to secure its supposed wealth: 'This is the greatest real estate venture since the Pope carved up the New World' (p. 133). In 'The Long Watch' (1948a) the 'Moon Base' is a military nuclear weapons site: 'a plain studded with projectile rockets, like candles on a cake' (p. 268). 'Gentlemen, Be Seated' (1948b) is a classic Heinlein tale in this mode, a tall story of 'sand-hogs' on the moon who, trapped in a transport tunnel out of Luna City, seal up pressure breaches with their buttock cheeks. In 'The Black Pits of Luna' (1947) a family trip to a glassy radioactive-blast crater (!) nearly ends in disaster when a child runs away. 'The Menace from Earth' (1957) is a teenage love story set in a Luna City grown so large there is a 'Bats' Cave' where humans can fly freely. "It's Great to be Back!" (1946) is possibly Heinlein's most seductive piece of propaganda for his inhabited moon. A couple of young lunar settlers flee back to Earth, only to run up against the ignorance and prejudice of the 'groundhogs', and such inconveniences as the heavy gravity, the filthy city air, and rampant viruses. They rediscover the joys of life on the moon: 'The fact that Luna City is the most comfortable environment man ever built for himself is beside the point—it's the people who count. Let's go home' (p. 318). Heinlein himself was the man who sold the moon, to his young readers. And it was Heinlein who would pen the most significant saga of lunar revolution (1966), as discussed in the next section.

The first lunar colonies, then, may be small establishments for specific purposes—science, military, resource utilisation—staffed by rotating shifts of specialists, and run on corporate, governmental or military lines. As islands of specialised labour surrounded by a lethal environment they might be less like small towns than like modern-day Antarctic bases, or deep sea oil rigs. Given automation, individual colonies may have very few personnel.

But it is from this unpromising soil that an independent lunar nation must grow. Clarke's *A Fall of Moondust* (1961), a suspenseful technothriller about the rescue of a 'Dust-cruiser' full of tourists crossing the Sea of Rains, is set on a moon evidently in political transition. As the presence of tourists from Earth suggests, this is a moon with a number of colonies large enough to be called 'cities,' including the largest of all, Clavius City, with over 52,000 inhabitants (p. 72). The lunar-born know nothing of Earth history before the first lunar landings (p. 133)—and yet the government of this young world is no more than a bureaucracy reporting to Earth-based authorities, under a 'Chief Administrator', just like the smaller science and industry bases of earlier times. One can imagine growing tension between the increasingly independent lunar dwellers and the rigid, outdated, unaccountable bureaucracy that rules them.

6.4 Liberty on the Moon: Essential Services, Governance and Policing

The most significant legacy of Breuer and Williamson's 'Birth of a New Republic' (1931) is probably its influence on Robert Heinlein who (according to an introduction to the 1981 edition by Williamson) acknowledged its influence on Heinlein's own much more substantial *The Moon is a Harsh Mistress* (1966). In 2076 Luna City is a penal colony. The convicts make a living by mining lunar water and growing wheat, which is exported—using a 'catapult', a mass driver—to an overpopulated Earth run by the 'Federated Nations', a stronger version of the UN. A Lunar Authority, under a Warden, controls the colonists' lives, including the provision of the air they breathe. The demands of the Earth-based authorities are excessive to the point that they will lead to famine on the moon in a few years' time. The lunar colonists rebel, and ultimately the confrontation is resolved as the colonists use mass drivers to drop rocks on Earth cities.

The immediate challenge faced by Heinlein's rebels is their utter dependence on communal life support systems. They understand that it is suicidal to strike against the infrastructure itself: 'The woman had been in The Rock almost all her life … yet could think of something as new-choomish as wrecking engineering controls' (p. 44). Meanwhile the Warden controls these essential systems centrally, from his isolated and heavily guarded complex. In the end Heinlein resolves these problems rather easily by giving the colonists a crucial ally in 'Mike', the colony's sentient central computer, which enables them to disable the Warden's Complex by having Mike subvert *its* oxygen supply. This does illustrate however the significance of the control of central life-supporting functions to the issue of liberty on the moon.

Since humans will expire in seconds if deprived of oxygen, the supply of air in an enclosed environment has an importance qualitatively different from other resources on which we depend, and for which even in the modern world we are prepared to pay—power, food, even water—and is central to the question of extraterrestrial liberty. Perhaps the most familiar dramatisation of this is the Arnold Schwarzenegger movie Total Recall (1990) which relocates some aspects of Heinlein's lunar revolution to Mars. A 'Mars Federal Colony' is ruled by Administrator Vilos Cohaagen, whose only priority is the continued supply of minerals from the mines of Mars, and whose use of the air supply ensures complete control: 'If you want to breathe you have to buy his air.' When revolution foments in a red-light district called Venusville, Cohaagen has no hesitation in cutting the air supply altogether, accompanied by memorable images of duct fans spinning to a stop. In the end Schwarzenegger's memory-addled hero Douglas Quaid saves the day by starting up alien atmosphere-generating machines.

Looking further ahead, what happens after the revolution? On the moon (as on other colony worlds) all the necessities of life will have to be provided by artificial means, and a key economic and social issue for a nascent lunar state will be the provision of these services. On an independent moon, how are the inhabitants to be provisioned with the essentials for life, day to day? Are air, water, power to be provided free? Or perhaps free at the point of delivery, paid for through a 'national insurance' taxation? Or are they to be provided only on presentation of direct payment, by those who can afford it? And if the latter, what is to be done with those who can't afford it? What guarantees can be given that the dependence on such systems will not be used as a lever for tyranny?

Heinlein dramatises one answer to these questions. The subsequent development of Heinlein's independent Luna is glimpsed in *The Cat Who Walks Through Walls* (1985), published two decades after *The Moon is a Harsh Mistress*. This is a fantastical tale of multiple realities, linking together many of Heinlein's earlier fictions. But there are long sequences set in a 'Luna Free State' (p. 61) over a century after the revolution (p. 180). Luna is presented as an arena of ferocious, unregulated and uncontrolled *laissez-faire* capitalism; common systems like a 'ballistic tube' transport system are 'privately owned and totally unregulated' (p. 185). And you have to pay directly for the air that you breathe (p. 190). It is the responsible of individual citizens to preserve their own lives; the state is not to be relied on, even for the most basic essentials. Just as in *Harsh Mistress*, this is underpinned by a harsh morality based on the dictates of survival, and a justice administered by the citizenry: 'Nobody ever touches another Loonie's p-suit [pressure suit]. Automatic elimination at the nearest airlock. No excuses' (p. 135). Heinlein's prejudices against government of any kind are clearly expressed: 'He thinks the world owes him a living … He thinks of "the government" the way a savage thinks of idols' (pp. 197–198). But such government as does function is shown as dictatorial and inefficient. One very elderly survivor of the revolution complains, 'When this was a penal colony, there was more freedom under the Warden than there is now with self-government' (p. 236).

A consciously different answer to the problem of essential-services provision is dramatised in John Varley's *Steel Beach* (1992). This is a knowing revisiting of Heinlein's ideas—indeed there is a cult of hard-line 'Heinleiners' (p. 50) on

Varley's colonised moon, many of whom take names from Heinlein's fictions (p. 387). In the 23rd century, mankind has been expelled from Earth by aliens known as the 'Invaders', but survives on the moon and other worlds of the solar system. While there are 'Disneyland' simulations of Earth landscapes and ecologies into which some escape, after 200 years people are acculturated to the lunar environment. The 'steel beach' of the title is the moon, onto which we crawled from the lost Earth and have begun to evolve, like lungfish (p. 294).

Varley depicts a nascent civilisation suffused by reasonably high technology, and this rich society has moved on significantly from Heinlein's frontier-austerity toughness. Despite widespread automation, a 'job guarantee is a civil right' (p. 146). But on the other hand, 'if you don't want to work, that's fine, too. Nobody starves, and air has been free since before I was born. It didn't used to be that way. Right after the Invasion if you didn't pay your air tax, you could be shown to the airlock without your suit. I like the new way better' (p. 147). Economically this system, of free access to essentials, is close to the post-scarcity model of economics (Chernomas 1984) (and is close to the constitutional model Robinson described emerging on his free Mars (1993, 1994, 1996)). The rational-anarchist Heinleiners however believe mankind has become soft (p. 50). Indeed, in *The Moon is a Harsh Mistress*, the idea of providing security of food, water, air and other essentials to the lunar citizens is raised but mockingly dismissed (1966, p. 159).

The debate here is about the extent to which the government of a lunar colony has a duty to preserve the lives of its citizens, in terms of maintenance of physical shelter and provision of essential supplies, in particular air, but also water, heating, food, security, and other necessities. The alternative is to regard it as the duty and responsibility of private citizens to provide their own and their families' air and other essentials. Heinlein's and Varley's visions of moon colonies offer diametrically opposed solutions to this debate, but of course a spectrum of constitutional and economic arrangements is available.

Policing as a challenge to the maintenance of human rights will be another issue in enclosed extraterrestrial environments. In Heinlein's moon, as depicted in *The Cat Who Walks Through Walls* (1985), the only policing is quasi-private—a settlement called Hong Kong Luna has 'Vigilante Volunteers' whose services you pay for (p. 156)—but even so, Heinlein's viewpoint character objects to their intrusion: 'I would not remain in a city that had sunk so far in the qualities that constitute civilisation that a cop would break into a citizen's bedroom merely because some officious official decides to summon him … A private soldier … has more freedom and more privacy than that' (p. 171). As a result Luna is a violent and weapons-rich place. We witness bandit attacks on the surface (p. 147ff) and gun-fights inside habitats (p. 243ff), as well as threats of more violent actions. The combination of violence, arms, poor policing and a fragile infrastructure would mean that such a community would always live in fear of imminent destruction, even though individuals have a certain freedom.

By contrast freedom on the moon of Varley's *Steel Beach* (1992) is heavily supported by technology. People live in a 'free society' with a government that is 'small and weak', but only because 'many of the instrumentalities of repression have been gradually given over to machines—to the Central Computer—not without initial trepidation, and not without elaborate safeguards' (p. 101). And in a further conscious rejection of Heinlein's values the days of the US gun lobby are remembered in bemusement: 'It was a civil right, the right to own firearms … But I'd have owned one, if I'd lived there. In an armed society, the unarmed man must be a pretty nervous fellow' (p. 51).

A lunar colony will be necessarily high-tech, and this may offer solutions to the policing dilemma. While abandoning responsibility to a central computer may not be wise (and indeed things go horribly wrong in Varley's novel), in enclosed habitats intensive, even total surveillance may be possible as a way to underpin effective policing. Of course this can be another route to tyranny. But Brin, in *The Transparent Society* (1998), has argued that an effective counter to ensure liberty is to have the watchers watched: to allow every citizen scrutiny of the surveillance process, and thus to exert control. (This has been explored in fiction, for instance in the author's own *The Light of Other Days* (2000, with Clarke and Baxter).

6.5 Liberty on the Moon: Local Resources

The economic future of a lunar nation will depend on the resources of the moon. In the short term mere survival may depend for example on the use of water from polar ice traps, and minerals for the manufacture of such requirements as solar power cells, habitat structures and even oxygen. In the longer term the physical nation's expansion must depend on the import of life-supporting volatiles, which the moon lacks. These will be purchased presumably through the sale of local resources, and so it seems likely that a moon colony will always be a trading nation.

According to the UN Outer Space Treaty (OST) of 1967 and its lunar accords, the moon and its natural resources are not subject to national appropriation. Presumably by the time of scenarios like Steele's (1991) and Bova's (1996), rights of ownership and exploitation of extraterrestrial resources will have been defined in practice. But any such treaties, like the OST itself, would however presumably be violated by the emergence of a nation *on* the moon itself.

Indeed, as the moon is likely to be the first colony world to attempt independence, the very lack of precedents may set unique legalistic challenges. In MacBride Allen's *Farside Cannon* (1988), set in the year 2116 (p. 56), there are political and economic tensions between Earth and the 'Settlement Coalition' (p. 27), the colonists of solar system bodies. MacBride Allen sketches (p. 134ff, p. 246) a law and custom of ownership of the territory and resources of celestial bodies, as deriving from terrestrial precedents such as historic European colonisation and the 1960s outer space treaties, and then (in the novel) refined by the field experience of actual colonisation.

But these policies were applied only imperfectly in the case of MacBride Allen's moon, precisely because it was the first body to be colonised. As a result, the 22nd-century moon is run by to a 'UN Lunar Administration Council' (p. 27), and in practice 'the moon is just another UN operations site, no more needing or deserving of its own currency than … an oil rig' (p. 153). The result is inefficiency, corruption, graft, and a total lack of democratic accountability; the most basic human rights are commodities to be purchased (p. 142). And meanwhile the colonists of the Settlement look down on the moon as a 'miserable eunuch colony' (p. 43).

In the aftermath of a move for independence, then, it may be necessary to establish national ownership of physical resources such as minerals, lunar ice and helium-3; perhaps a corporation or corporations will be established to exploit those resources in accordance with terrestrial law. All these moves may require the renegotiation, or rejection, of existing treaties.

6.6 Liberty on the Moon: External Resources and External Relations

As mentioned in the previous section, one brutal fact of lunar existence will be a reliance on the import of resources from beyond the moon for the essentials of life. The near-future moon, deficient in the requirements to support human life, even in water, can never be environmentally independent of the rest of the solar system, whatever its political autonomy.

In Bova's *Moonwar* (1997), Moonbase attempts to declare its independence. But self-sufficiency from Earth is presented as a prerequisite: 'But you are not a nation!' says the (corrupt) UN Secretary General. 'Two thousand people do not make a nation! You can't even exist by yourselves without supplies from Earth. It is as if a group of people on an ocean liner declared themselves an independent nation. It is nonsense!' (p. 82). Similarly a reliance on extralunar resources for further growth is pointed out in Steele's *Lunar Descent* (1991), in which the newly independent colonists have plans to bring down water from a near-Earth asteroid.

Given this dependence on external supplies, a significant constraint on the development of any independent lunar state will be its relationship with Earth and its nations and corporations. And because Earth is so physically close by, the home planet will surely remain economically and militarily dominant over lunar colonies, even compared to Earth's relationship with extraterrestrial colonies further out in the solar system.

Imperial Earth (1975), Clarke's elegant saga of interplanetary politics set in the US Revolution quincentenary year of 2276, is predicated on an interesting observation about the human effect of physical proximity: 'The global electronic village which had existed for centuries on the mother world could never be extended into space; the political and psychological effects of this were enormous, and still not fully understood … The communications satellites had made possible, and then inevitable, the creation of the World State in all but name … There were perhaps a thousand key individuals … and they talked to each other incessantly from pole to pole … But with the establishment of the first Mars Base, that intimacy had ended.

Earth could talk to Mars—but its words would always take at least 3 min to get there' (pp. 38–39). Knowing this, senior figures from colony worlds are forced to travel to Earth to establish personal contacts. And, though it is not stated explicitly, the moon, only a few light-seconds from Earth, and presumably colonised long before Mars, is drawn into the intimacy of direct contact with Earth's inhabitants: the moon is seen an appendage of the Earth.

This perspective is reflected in others of Clarke's fictions, such as the last book of the 'Time Odyssey' sequence by Clarke with the author (Clarke and Baxter 2008). A tentatively colonised moon is central to efforts to save the Earth from alien intervention, but the 'Spacer' communities 'don't find the Moon very interesting. They aren't true Spacers down there. Not when you can commute to Earth in a day or two. We call it Earth's attic …' (p. 74).

A similar depiction of a politically compromised moon is Niven's *The Patchwork Girl* (1980). In the year 2126 (p. 11) humanity is divided into three: 'flatlanders', Earth dwellers under a UN government; 'Lunies' on the moon; and 'Belters', pioneers in the asteroid belt. The story is an ingenious murder mystery. As humanity diverges—'we were three branches of humanity, trooping towards the lights' (p. 11)—there are wide ethical differences between Earth and Belt, especially over the issue of capital punishment. On an overpopulated Earth the bodies of the convicted are harvested for their organs. The Belters recoil from this, but their compromise is to send their own condemned to the moon for execution, a solution that makes nobody happy. The Lunies feel constrained: 'The moon is a kind of halfway house between the mud and the sky … between Earth and the Belt. We gain some advantages from that, but we have to keep you both satisfied too' (p. 48).

Niven also depicts interesting demographical divergences from Earth. An overcrowded Earth is stultified by population-control laws (and the crime in the story is motivated by an attempt to get around those laws). But on the moon, while fertility is difficult in low gravity, the population is growing fast (p. 38). A fecund people dominated by the young would have many ideological differences with an Earthbound gerontocracy.

6.7 Discussion

The picture that emerges from science fictional depictions of the governance of an independent lunar nation is a complex and detailed one. Indeed, even juvenile portrayals of lunar colonies can hint at political complications. In Gerry Anderson's *Captain Scarlet and the Mysterons* (1967), in the year 2068 the moon is caught in the crossfire of an alien war. In the episode 'Lunarville 7', the 'lunar controller' declares that 'The Earth is engaged in a savage war of nerves with the Mysterons [on Mars]. This is none of our doing and we want no part of it.' The moon is to be an 'independent world' and the lunar dwellers will step aside from the war. Captain Scarlet discovers that, alas, the controller has been taken over by the Mysterons, who are building a city of their own on the moon.

Before independence, an inhabited moon is likely to host many colonies, sponsored by different nations, corporations and other bodies, and serving many purposes: military, scientific, commercial. Such bases will have their own separate cultural origins and traditions and must necessarily be self-reliant at least in terms of short-term emergencies. Some colonies may not wish to have any part in an independence movement—such as the multinational scientific staff of a lunar surface observatory caught in the crossfire of interplanetary war in Clarke's *Earthlight* (1955)—and decisions may need to be made whether to expel or accommodate such communities. On the other hand the newly independent colonies may cooperate and share common facilities such as transport systems or off-moon launch systems like a mass driver, and they may have common interests such as in science reserves. A moon of many cultures may be part of the background to independence struggles.

A governance system to manage a nucleated human geography must therefore be devised. The total population of the moon may not be large, and individual bases may be small communities compared to modern-day terrestrial towns and cities. The nuclear colonies themselves may have governance systems suitable for small populations, for example fully participatory democracies. A federal system may be drawn up to incorporate semi-independent 'city states', largely self-governing but with shared interests, such as common transport systems, mutual aid, and relations with external bodies such as Earth governments. A lunar society might be similar in some regards to the leagues of Greek city-states of antiquity.

In an environment where all human life depends on the operation of machinery, early decisions must be made about the provision of life support, most crucially air to breathe. At one extreme total provision of life's necessities could be provided free of charge; at the other extreme every instance of supply might be individually charged for. A spectrum of compromise arrangements is of course available. Safeguards must be inbuilt to ensure that it is impossible to use such systems to limit the liberty of the population.

The economic future of a lunar nation may depend on the resources of the moon, for survival in the short term and as a basis for off-world trade and growth in the long term. It may be necessary to establish national ownership of physical resources. All these moves may require the renegotiation, or rejection, of existing treaties. Considerations will therefore be necessary as to the economic status of the new nation. It may declare ownership of the whole moon and all its resources; it may exploit those resources itself or license others to do so. It may establish itself legally as a single corporation under the laws of Earth or a nation thereof, as depicted by Breuer and Williamson (1931) and Steele (1991).

The scientific value of the moon represents another resource. Lastra (2014) has noted that 'The primary scientific importance of the Moon arises from the fact that it has an extremely ancient surface (mostly older than 3.5 billion years, with some areas extending almost all the way back to the origin of the Moon 4.5 billion years ago)' (p. 81). Thus the moon might contain clues to the origin of the terrestrial planets, to the environment of the early solar system, perhaps even the conditions that led to the origin of life. This environment could be valuable as a resource for lunar colonists, if only for the income to be generated by supporting visiting scientific staff. Yet this treasure, dependent on the stillness of the moon and its undisturbed nature, would be disrupted by human colonisation and associated activities, notably large scale resource extraction. Perhaps a priority for an independent lunar government will be the establishment of zones of environmental protection, and/or to establish an 'environmental court' to argue the case for the landscape against competing interests, as suggested by Robinson in the analogous situation of an independent Mars (1996).

The moon is likely to be an anomaly in the political history of a colonised solar system, precisely because it will most likely be the first world colonised. It will be a test bed for political, economic and other arrangements, including the renegotiation of treaties concerning local resource exploitation. Lessons from mistakes made there may benefit the colonists of worlds further out, but the moon itself may be left with a mish-mash of flawed constitutional arrangements.

One long-term interest for all lunar colonists will be political relations with Earth and the wider solar system. While aspects of lunar life such as confinement and the reliance on communal systems will be in common with settlements on other worlds such as Mars, the moon's relationship with Earth will be unique. A lunar society is always liable to be dominated economically, and perhaps politically and psychologically, by its giant neighbour the Earth. While the lunar colonists may feel more akin culturally to the colonists of worlds further out, from the point of view of those worlds, from which Earth and moon will appear together as a mere double star, the lunar colonists may appear lesser creatures forever in Earth's shadow.

Lunar colonists may make decisions about their political identity based on their perception of these external relationships. They could constitute as an entity acceptable to join existing Earth governance systems, such as a state to join the United States of America—Hawaii is a precedent of a physically remote territory joining the union—or a nation state eligible to join the United Nations. Perhaps it could become a province of a future World Government. If an extraterrestrial federation already exists (as suggested by Crawford (2014) as a hedge against the dire possibility of interplanetary war), the moon might apply to join—or, if the moon is the first extraterrestrial nation, it might create such a federation.

Certainly a lunar society will be one of intersections, between Earth and the frontier worlds, between past and future. The establishment of an equable system of governance for its colonists may present unique challenges.

But the heart of any new constitution must be a vision of a better society. Clarke argued that the moon itself has been central to our perception of the world since before we were fully conscious. The first character we meet in *2001: A Space Odyssey* (1968), three million years before the present, was a 'man-ape' called Moon-Watcher: 'Of all the creatures who had yet walked the Earth, the man-apes were the first to look steadfastly at the moon' (p. 8). Now we will begin to live there.

And the moon itself may prove to be a stepping stone to the future in a way we have not anticipated. To quote fiction by a man who has been there (Aldrin and Barnes 1996): 'For the first time, what he had dreamed of since he was a small boy had happened: his boots were planted on the soil of another world, and when he looked up he saw, not home, but only stars. He might have been anywhere in the universe at that moment, for every galaxy must be full of small, stony worlds …' (p. 123).

If we can learn to build a functioning human society on the moon, we can build such a society anywhere.

References

Aldiss, B. (1962). *Hothouse.* London: Faber and Faber. Page numbers from the 1979 Panther edition.

Aldiss, B. (1986). *Trillion year spree*. London: Gollancz.

Aldrin, B., & Barnes, J. (1996). *Encounter with Tiber.* New York: Warner.

Baxter, S. (1998). *Moonseed.* London: HarperCollins.

Baxter, S. (2001). *Omegatropic*. London: British Science Fiction Association.

Brin, D. (1998). *The transparent society*. New York: Perseus.

Bova, B. (1996). *Moonrise.* New York: Avon. Page numbers from the 1997 New English Library paperback edition.

Bova, B. (1997). *Moonwar.* New York: Avon. Page numbers from the 1997 New English Library paperback edition.

Bova, B. (2013). *Farside.* New York: Tor Books.

Breuer, M. J. & Williamson. J. (1931). *The birth of a new republic.* New York: Amazing Stories Quarterly dated winter 1930. Page numbers from the 1981 reprint by PDA Enterprises, New Orleans.

Burroughs, E. R. (1926). *The moon maid.* New York: A.C. McLurg.

Captain Scarlet and the Mysterons. (1967). *Century 21 Productions.* 'Lunarville 7' written by T. Barwick.

Chernomas, R. (1984). Keynes on post-scarcity society. *Journal of Economic Issues, 18*(4), 1007–1026.

Chilton, C. (1954). *Journey into space.* London: Herbert Jenkins.

Clarke, A. C. (1953). *Prelude to space.* London: Sidgwick and Jackson.

Clarke, A. C. (1955). *Earthlight.* London: Frederick Muller. Page numbers from the 1957 Pan edition.

Clarke, A. C. (1961). *A fall of moondust.* London: Gollancz. Page numbers from the 1964 Pan edition.

Clarke, A. C. (1968). *2001: A space odyssey.* London: Hutchinson. Page numbers from the 2000 Orbit edition.

Clarke, A. C. (1975). *Imperial Earth.* London: Gollancz. Page numbers from the 1988 VGSF edition.

Clarke, A. C., & Baxter, S. (2000). *The light of other days.* London: HarperCollins.

Clarke, A. C., & Baxter, S. (2008). *Firstborn.* London: Gollancz.

Clute, J. et al. (2014). The encyclopaedia of science fiction. www.sf-encyclopedia.com.

Cockell, C. S. (2014). *The meaning of liberty beyond Earth.* Switzerland: Springer International Publishing.

Crawford, I. A. (2014). Interplanetary federalism: Maximising the chances of extraterrestrial peace, diversity and liberty. In Cockell (2014).

Doctor Who, BBC: 'The Moonbase' by Kit Pedler, broadcast 1967. Novelisation by Gerry Davis. 1974. London: Target Books. 'The Invasion' by Derrick Sherwin, broadcast 1968. 'The Seeds

of Death' by Brian Hayles, broadcast 1969. 'Frontier in Space' by Malcolm Hulke, broadcast 1973. 'Victory of the Daleks' by Mark Gatiss, broadcast 2010.
Gribbin, J., & Chown, M. (1988). *Double planet*. London: Gollancz.
Griffiths, G. (1900). *Honeymoon in space*. London: Pearson.
Heinlein, R. A. 1939 ('Requiem'). 1941 ('Logic of Empire'). 1946 ('It's Great To Be Back!'). 1947 ('The Black Pits of Luna'). 1948a ('The Long Watch'). 1948b ('Gentlemen Be Seated'). 1949 ('The Man Who Sold the Moon'). 1957 ('The Menace from Earth'). Compilation: Heinlein, R.A. (1977). *The past through tomorrow*. New York: Berkley Medallion. Page numbers from the two-volume 1977 NEL edition.
Heinlein, R. A. (1947). *Rocket ship Galileo*. New York: Scribners.
Heinlein, R. A. (1966). *The moon is a harsh mistress*. New York: Putnam. Page numbers from the 1969 NEL paperback edition.
Heinlein, R. A. (1985). *The cat who walks through walls*. New York: Putnam. Page numbers from the 1988 Ace edition.
Howe, D. J., & Walker, S. J. (2003). *The television companion: The unofficial and unauthorised guide to doctor who*. Tolworth, England: Telos.
Iron Sky (2012). Walt Disney, dir. Timo Vuorensola.
Johnson, R. D., & Holbrow, C. (Eds.). (1977). *Space settlements: A design study*. Washington: NASA SP-413.
Lastra, C. (Ed.). (2014). *Manifesto for a republic of the moon*. London: The Arts Catalyst, Aldgaye Press.
Leinster, M. (1957). *City on the moon*. New York: Ace.
Lewis, C. S. (1945). *That hideous strength*. London: Bodley Head.
MacBride Allen, R. (1988). *Farside Cannon*. New York: Baen.
Moon (2009). Sony Pictures Classics, dir. Duncan Jones.
Niven, L. (1980). *The patchwork girl*. New York: Ace. Page numbers from the 1982 Futura edition.
O'Neill, G. K. (1976). *The high frontier*. New York: William Morrow.
Roberts, A. (2006). *The history of science fiction*. London: Palgrave.
Robinson, K. S. (1988). *The lunatics*. New York: Asimov's Science Fiction Magazine.
Robinson, K. S. (1993). *Red Mars*. New York: Random House.
Robinson, K. S. (1994). *Green Mars*. New York: Random House.
Robinson, K. S. (1996). *Blue Mars*. New York: Random House.
Ryan, C. (Ed.). (1952). *Across the space frontier*. New York: Viking Press, 1952. Page numbers from the 1952 Sidgwick and Jackson edition.
Scheer, K. H., & Ernsting, W. (1961). *Perry Rhodan #1: Enterprise stardust*. Germany: VPM. Page numbers from the 1969 Ace translation.
Slyuta, E. N., Abdrakhimov, A. M., & Galimov, E. M. (2007). *The estimation of helium-3 probable reserves in lunar regolith*. Lunar and Planetary Science, XXXVIII.
Spudis, P. (1997). *The once and future moon*. Washington: Smithsonian.
Steele, A. (1991). *Lunar descent*. New York: Ace. Page numbers from the 1992 Legend edition, London.
Steele, A. (1996). *The tranquillity alternative*. New York: Ace.
Steele, A. (2014). *V-S day*. New York: Berkley.
Tcherniavski, I. (2014). Possibility of detection of Earth-like exo-planets and recognition of their surface areas with a hypertelescope. *Optical Engineering*, *53*(2), 023106.
Total Recall (1990). Tristar Productions, dir. Paul Verhoeven.
UFO (1969). Century 21 Productions: 'The Dalotek Affair', written by R. Powell. 'The Responsibility Seat', written by T. Barwick.
Varley, J. (1992). *Steel beach*. New York: Putnam. Page numbers from the 1993 HarperCollins edition.
Verne, J. (1870). *Around the moon*. Paris: Hetzel.
Wells, H. G. (1901). *The first men in the moon*. London: Newnes.

Chapter 7
The Marxist Theory of Capital as Applied to the Economic Development and Governance of Space Colonies

John R. Cain

Abstract The results of living and working extra-terrestrially are not yet known but it is expected that because of the need to survive in the hostile environments of space that the economic and political structures of the settlements will be socialistic/collectivist in nature. A study of Marx's theory of capital, therefore, should provide an understanding of the economic and political workings of the space settlements as they develop. This chapter outlines Marx's ideas on production, on the money supply, on profit, on labour and on the origin of economic crises as they are detailed in his major work "Capital: Critique of Political Economy". It then examines how these ideas relate to and may function in an extra-terrestrial society.

Keywords Marx's "Capital" · Extra-terrestrial economics · Production · Money supply · Labour · Space entrepreneurs · Economic crises

7.1 Introduction

With the developments in modern science and technology and their application to improve mankind, we are moving from the idealised practical everyday life of Man that we conceptualise to one that is practical but radical in its thinking. Mankind for the first time in its history can think of space as an environment that needs to be conquered and there is now the knowledge and ability to do so (Sawyer 2012). This change in thinking is revolutionary and will have profound affects on how Man projects the vision of settling other worlds.

In the early stages of the colonisation of other worlds, capital will still be needed to finance projects and to guarantee that the necessary goods reach the settlements. Because of the nature of capitalism and its changing face as seen after the world financial crisis of 2008, the theories of Marx as they are outlined in his major work

J.R. Cain (✉)
GeoFind Consultancy, Hookstone Chase, Harrogate HG2 7HN, North Yorkshire
e-mail: johncain6@hotmail.com

C.S. Cockell (ed.), *Human Governance Beyond Earth*,
Space and Society, DOI 10.1007/978-3-319-18063-2_7

"Capital: Critique of Political Economy—Volumes 1–3" (1990, 1992a, b) are being re-interpreted as a means of understanding this change (Perelman 2010). This is important as the form of society that may respond best to Marxist doctrine may be extra-terrestrial in nature. Cockell (2013) has stated that "……there are several aspects of a Marxian society the potential appeal of which is easily identified" and includes for example, the exact control of the supply of commodities, the lack of private property and the sense of community.

Marx's "Capital" was published in three volumes i.e. in 1867, 1885 and 1894. Volume 1 critically analyses the contradictions of the capitalist mode of production including the concept of the class struggle. Volume 2 examines the circulation of capital within the financial system including the owners of money and surplus value. Volume 3 analyses the process of capitalist production including an examination of the value of interest and profit and loss. A forth volume published after Marx's death in 1883 examines surplus value theory (Marx 1975).

The initial stages of space colonisation will eventually lead to the evolution of specific economic applications and related governance as more of the Solar System is settled and extra-terrestrial societies struggle in a hostile environment to survive (Cockell 2013). Because of the innate socialistic structures of an extra-terrestrial society (Merrifield 2013), a study of the theory of capital as postulated by Marx should give us an understanding of the economic workings of the early space settlements and how this may affect the governance of such a society in particular in an economic crisis. Marx together with his friend Engels (2011) were both well aware of the transitionary nature of history and how events could lead to major changes both in terms of how capital was distributed and how peoples lives were governed.

This chapter aims to analyse Marx's broad theory of capital as detailed in the four volumes of "Capital" and how it may be relevant to the establishment, governance and sustainability of "space settlements". In the early stages of space colonisation, there is likely to be what will be an "imperialistic" stage. There will also be power struggles between astronauts and other space explorers for resources and prestige fostering the development of societies that could appear tyrannical in nature (Cockell 2008, 2009). The use and function of capital and how it relates to those living and working in space therefore will be especially important.

7.2 Forms of Production

Marx postulates in Volume 2 of "Capital" (1992a) that there are two forms of production i.e. "simple reproduction" where production is renewed at the same level as previously and "expanded reproduction" where surplus-product is used to increase production. In the former, the economy is more likely to stagnate than grow whereas in the latter, production will increase. In some settlements that have been in existence for many years, it is expected that there will be sufficient consumer goods produced to feed the needs of the astronauts and other workers adequately namely “simple reproduction”. To generate “expanded reproduction” there will be a need for investment in sufficient machinery, raw materials and other components. This could lead to a demand for more labour.

In “Capital” Volume 2 Marx (1992a) saw a distinction between two broad economic sectors which he named “Department 1” and “Department 11”. Department 1 includes the means of production for example, in space the factory provision of oxygen, the mining for rare asteroid ores. Department 11 describes the production of consumer goods such as clothing and in space this would include spacesuits and the food produced using hydroponics inside a settlement or spaceship. The scale of these two economies will fluctuate considerably as a settlement expands and extends its reach. Department 1 industries will become very large scale when utilised extra-terrestrially for example, in the use of solar power to generate energy. For both Department types to function, they must provide the goods in specific proportions otherwise there may be a surplus in one Department type (“extended reproduction”) and scarcity in another (“simple reproduction”).

7.3 Commodity Value

Basic economic theory believes that any labour market supplies a variety of commodities that a population wants or thinks that it needs hence the use of marketing principles to try and sell the products. The value created by any exchange will be represented in the products sold (Bellofiore 1989). In space settlement communities, specific items will be invaluable to enhance survival. Such items will have use value that will be decided by the customer but linked to the price, packaging and promotion of these goods. According to Marx (1975), there is a distinction that needs to be made between the use-value of a commodity and its exchange value. A use value must meet a specific human need such as providing effective communications systems in space whereas the exchange value reflects the commonality between the goods.

The value of a good reflects its exchange value and this in turn reflects the cost to society and the labour power involved in producing the commodity (Perelman 1993). Where survival in a space settlement is concerned, the cost to supply the goods to society will be as important. However, even extra-terrestrially, it is expected that certain items such as works of art will command a higher cost but only if there is time for individuals or the community to enjoy them. Money spent on such commodities may give a false impression of their worth.

Because of the skills, knowledge and intelligence required to train an astronaut, they will attract a high salary. It is the value placed on them by an employer that will determine their salaries. Marx was well aware of the principal of value and in “Capital”—Volume 1 (1990) gave an example of share prices that fluctuate i.e. they rise in a bull market and drop in a bear market. Such rises and falls depend on what customers are willing to pay for the shares. In this respect the market cannot be predicted but hedge funds and the predicting of future commodity values try to beat the market trend. Hedge funds generate greed and the slow undermining of capital. This form of sophisticated betting, if practiced by a few individuals, could have calamitous consequences (e.g. destabilise the government) if practiced in a space settlement. Safeguards would need to be implemented to prevent this occurring. However such safeguards could upset market prices and reduce investment within a particular colony resulting in stagnation. Investing in the market therefore could affect the two-fold character of labour such as its cost and contradict Ricardo's concern with the "magnitudes of value" (Sraffa 1951) and the ways of predicting market processes.

In almost all societies, capitalism has been shown to be the motor of change. Without its power to invent and provide society with its needs, there would be a reverting back to some form of a feudal system as in mediaeval times. In the early stages of space colonisation, it is highly probable that extra-terrestrial societies will use capital to produce, sustain and develop those commodities that it needs to survive.

As the returns from mining minerals and the setting up of factories and laboratories in settlements generate profits, more private revenue will be pumped into space exploration. In these situations however, there will be demands on the terrestrial capitalist system to provide the funding with each settlement being in direct competition with each other for the finance and necessary human resource. The values that could lead to competition for funding between settlements may not be as aggressive as on Earth. A new economics based on the need to provide a more fluid movement of money may become established. It will be based on individual patterns of space utilisation and entail closer relations with the financiers to provide capital for those commodities absolutely necessary for survival such as oxygen and water.

In space settlements as on Earth market prices to some extent will dictate the value of the goods being provided. However, in space there will not be the same need for luxuries as on Earth because of the lack of personal living space (Daily and Ehrlich 1992). Until they can be produced cheaply by the settlers themselves, the luxury commodities will be costly due to the high costs of transporting them from Earth. The providers therefore will be at an advantage because they will have monopoly over supply and be in a position that will be open to abuse.

7.4 Products of Capital

The "products of capital" as defined by Marx in "Capital"—Volume 1 (1990) are simply what the capitalist requires. This is not only a return on the variable capital invested in space projects for example, but rather in the total investment i.e. variable capital plus the constant capital linked to the means of production. When companies invest to mine precious ores from asteroids or from Mars and the Moon, the profit motive will be uppermost because capitalism will dominate. But there will be "altruistic investment" to develop the commodities necessary for survival such as oxygen, efficient waste removal systems and food processing. It is expected that in most instances such services will not be so dependant on the profit motive unless there are many providers of such services and there is competition between companies. In all circumstances, it will be those that exploit the markets most successfully who will reap the rewards. Labour costs for key workers will need to be high because of the technical nature of the exploration and exploitation of space. Nevertheless, the rate of profit generation will need to be such that there is continual investment and this may only be attainable by effective low labour costs (Jameson 1984).

The provision of suitable wages to match the job is likely to reflect Marx's concept of the "organic composition of capital" (1990) where there is a ratio of constant capital (for example, from that generated from the mining of ores) to the variable capital (for example, from that received from providing fresh oxygen). In value terms, the money generated to produce a commodity, will be relative to the labour power needed. This may result in stasis as the settlement expands. It remains to be seen whether the more efficient the labour power is that the productivity of the labour will increase and have a significant affect on the "organic composition of the capital". As the economics of space exploration develops, it will show the major changes in how capital is utilised on Earth and extra-terrestrially. There will need to be a trade off between the functions of the colony and those supplying the finance. Once this relationship starts to dismantle, then a new paradigm based on the colonists survival needs will be developed and implemented.

7.5 Price of Goods and Competitive Advantage

In the early stages of planetary and interstellar space exploitation, the price of goods will most likely be determined terrestrially by Governments and industrialists and not by the capitalist market system operating in space. It is expected therefore that the price of supplies will be cheaper than the market rate as the needs of the colonists will be uppermost. The practicalities of day-to-day living in space will ensure that the correct goods necessary for survival are purchased at the price as set by the settlement governing bodies. To gain a competitive advantage, some settlements may try to undercut prices and therefore it will be necessary for the "social labour" as envisaged by Marx to meet the conditions of production to keep prices low.

For example, space tourism will initially be costly and limited to a few people who can afford the financial outlay. The initial high flight costs will be because the technology required to produce a safe spacecraft and the skills needed to service and maintain the craft will be high. However, it is expected that costs will decline as more people wish to explore space and there is more competition to satisfy this human need. As space tourism develops and more space runways are built to cater for the demand, there will be intense competition between the suppliers of the service and the market share. Eventually these factors will reduce the costs of space travel overall resulting in an explosion of the exploration and settlement of both inner and outer space and the need for "workers" to service the tourists and adventurers. The development of space tourism will initially follow classic economics via the laws of supply and demand. But the struggle between suppliers of the space tourism business will need to walk side by side with the need to meet the demands of national prestige. Those countries that are not in the vanguard of providing space tourism will have lost out and thereby failed to gain competitive initiative.

7.6 Money Supply

The Marxist concept of the division of the capitalist economy is into separate, interdependent and competing producers together with the divisions within each unit of production. Such a division leads to a separation between capital and wage labour and should not initially be a major factor in extra-terrestrial economics (Moseley 1997). This is because the financing of the conquest of space will require money that is only going to be available from Government and wealthy individuals. The ordinary person could be excluded from the enterprise to conquer space unless for example, Governments float specific space adventures as rights issues and thereby give the public the chance to take part. The finance obtained will be used to employ a skilled labour force to establish and sustain the colonies. Huge financial investment will be needed in the early stages of space exploration and capitalism may not be able to provide it in the long-term unless there are continual high profit margins and increasing dividends for all those involved.

During the early stages of space exploration there will be the use of money to buy commodity items but this could change as new forms of exchange are developed. The formula $M\text{-}C\text{-}M^1$ has been devised by economists to explain where money is invested in order to produce commodities (C) which are then exchanged for money. The M^1 or money which an investor or capitalist holds following a transaction will be greater than M i.e. the money invested in the first instant. This extra value or surplus/profit could come from many transactions. In the early stages of colonisation because wages and salaries paid may not relate to profits in particular as most of the commodities for living will be provided e.g. oxygen, water, food, sanitation, life support maintenance, the scientists and engineers working to advance science and technology will have little cash to spend or invest. New working and reward systems will need to be developed to ensure that all those living and working together in a settlement (or world spaceship) including both the "skilled" and the "unskilled" labour forces remain content (Leysens et al. 2010).

Politicians and economists will be involved in devising social systems in which astronauts and others can work and develop as individuals (Przeworski and Limongi 1993). With careful management, the exploitation of sections of a settlement should not occur; the means of production and labour will be in the hands of the colonists themselves. But will the distribution of capital be fair within a space colony? Who will utilise the profits of an enterprise for example, from the mining of metals and ores on an asteroid? Will the profits be fed back into the settlement to stimulate growth? Will there be equality between the competing scientists and engineers? Who will undertake to provide the workers for the labour tasks that may be dangerous but need to be done? Will there be conflict between the two "classes" of worker i.e. the highly skilled and basic skilled or harmony as they try to achieve some forms of political and social stasis? These questions will be addressed below.

7.7 Money Movements and Capital Flows

Capitalism ensures that capital is accumulated and can then be invested. It is not possible to determine how this principal will be incorporated into the management of the settlements. But because of the nature of the society with its initial dependence on Earth to provide the essentials for life such as food, it is expected that a form of "bartering system" will develop where "money" is exchanged. If not, then all transactions may occur using personal accounts or even using a form of "bio-coins". A specific monetary transaction system used by a settlement will provide the best means of exchange for that settlement. Alternatively, there may be a system that encompasses an exchange in all its parameters but depends on cyber-net communications for transactions to succeed. Nevertheless, in any system used to ensure a smooth flow of money throughout a settlement there will still be a sub-population classed as wealthy and another as "poor" though not in the sense of how "poor" is defined terrestrially. Consumption of the products of production including the "use-value" will be identified by the colonists themselves and possibly via advice from financial managers on Earth As explained in "Capital"—Volume 2 (Marx 1992a) at some stage, any surplus value i.e. the capital accumulated, will be ploughed back into the production processes and produce more surplus value.

In a space settlement there will be capital flows between individuals and between specific groupings. Any profits generated are likely to be invested terrestrially as the returns will be higher, but over time some individuals together with successful companies will acquire extensive capital that they want to re-invest in the settlement for the benefit of the whole population. The drip-feeding of money into a settlement may be used to undertake specific projects. However, there will at some stage be a demand for luxury goods in the form of clothing, paintings, wine, books or as yet undetermined items. In addition, there will be a demand for the labour to produce the goods.

In the first decades of space colonisation, the resins, plastics and other materials required to produce the goods used by colonists such as 3D products will need to be supplied from Earth. Over time, those persons with the accumulated capital made through terrestrial/extra-terrestrial transactions will have leverage on the range and scope of any products manufactured and on the nature of the projects undertaken. This will give such individuals political influence as to how a settlement is organised or controlled. Such influence could have long-term damaging affects on a settlements growth.

Marx (1975) envisioned that the so-called "bourgeois" or if now applied to extra-terrestrial colonists as the "astronaut/technical" class would save as much of the surplus value as possible for reinvestment. In space, there will not be as great an opportunity to utilise the accumulated wealth. However, there may be the potential to invest the money on colonial endeavours through share options and bonds issued from Earth. These colonial bonds and shares are not likely to produce the same returns especially in the early stages of investment, and years of planetary exploitation may be needed (e.g. in the mining of precious metals) for a profitable return. However, human nature will provide the motive for investment even if the returns are low, namely greed, and the competition to become wealthier than the next person. The achievement of wealth will show in a form of self-aggrandisement and the form that this will take will be dictated by the extra-terrestrial environments in which the wealthy will live and work.

If not managed carefully, the spending to achieve self aggrandisement could lead to bankruptcy in particular if the accumulated wealth declines absolutely together with the surplus value of the goods bought. Bankruptcy may not be as damaging extra-terrestrially because the person(s) involved will still be a valued member of the settlement with the skills and experience needed to ensure the colony survives. But how will a settlement respond to near or full bankruptcy in important influential members of its governing body. Some individual colonies may not survive unless they are successful at re-investing any accumulated wealth back into the market and then onto the centres of production as part of a democratic process.

7.8 Exchange, Supply and Demand

The use of money and the payment of wages will still be required by the labour workforce to cater for their basic needs. For maximum efficiency in supply and demand, the amount of income given to the workforce will need to be regulated by the colony leaders. If the control of the money supply and its distribution was regulated from Earth, then there would be the potential for conflict as one side competed with the other on how it should be spent. In such circumstances the potential for conflict would necessitate the formation of a "planetary police force" to control any rebellions by the disadvantaged "labouring" classes.

The production and distribution of the money supply by individual settlements could lead to "settlement inflation", political divisions between groups and inter-colony rivalry for resources. To prevent such events, each colony could pool financial resources, utilise any monetary profit generated and use the money for the benefit of all the settlement population. In such circumstances, however, the profits generated would be low and accumulate slowly. To provide more money therefore, each settlement would have the choice to behave as a producer of capital and act accordingly or Earth financiers could intervene. Any restrictions on the money supply from Earth outwards would affect a commodities individual value, the labour time it embodies and its social or market value.

7.9 Profit and Investment—Space Entrepreneurs

Capitalists strive to gain the largest possible return from their investment. It is this motive that spurs entrepreneurs to take risks. During the early establishment of space settlements therefore, it is expected that there will be an equalisation of the rate of profit from other business ventures such as the provision of better means of surface transport, the designing of more comfortable spacesuits and the using of improved sampling techniques to assess health risks. The general rates of profit that entrepreneurs will generate will reflect the relationship between the total surplus value produced and the total social capital invested. This will ensure that a settlement survives as a profitable economic entity that will satisfy both private and public investors. However, what will be the outcome if two colonies are in competition with each other for limited resources and thereby generate unequal profits? Will some of the capital resource in terms of the scientific and the financial investment be transferred to the most successful colony leading the others to a downward spiral of poor investment and subsequent lack of progress? Or will those terrestrial financiers with investments in both these colonies try to equalise the situation so that a balance is restored? This restoration in the balance between the colonies may lead to increased production in both settlements that may continue until the supply exceeds the demand. It will be in the interests of all concerned both extra-terrestrially and terrestrially to see an equilibrium in the return of profits and in increased production as resources flow back into the economy. A failure to restore equilibrium could lead to a severe restriction on the liberty and freedom of the individual.

In economic terms, equilibrium will have been reached in an extra-terrestrial settlement when the prices of varying resources including the outcomes of scientific knowledge can be set at levels which earn relative profits. Due to having achieved equilibrium, there is likely to be an incessant outflow and influx of capital from and into a settlement similarly as Marx predicted for businesses during the nineteenth century (1975). A reduction in competition between settlements for capital and with the financial service providers will require skilled resource management to avoid a conflict for capital not only between the astronaut populations both terrestrially and extra-terrestrially but between other competing workers.

7.10 General Profit Rates and Social Implications for a Settlement

A rate of profit on the returns of investment for the provision of the space infrastructure will be an economic requirement. Marx (1992b) argued in Volume 3, Chap. 9 of "Capital" that the general rate of profit would necessitate the "transformation of values" into cost prices that would be different from the values (Perelman 1987). Each settlement will have prices of production because each "leader" including the astronauts and scientists will need to "exploit" the non-specialist workers. Any profits resulting from the exploration of space could be distributed only to the providers of the investment capital but this could generate unrest so those most influential in the settlements would probably benefit from the financial returns also. Space entrepreneurs will allocate a single market value for particular services and goods and this will be reflected in the single market price. This price will be estimated from the range of values of the commodities on offer. However, there will reach a stage when the sum of the prices of production of all commodities produced in a space settlement will be equal to the sum of all their values.

In space, social chaos could be the result of the unequal distribution of wealth unless it was well managed to result in a "transformation problem" with the general rate of profit utilised to ensure a modification of the law of values. Because of the structure of a settlement in terms of its physical nature and human resource distribution, it may be difficult to convert the prices of production for the values of the commodities being traded, including technical knowledge, from the variable and constant capital prices. Following any "transformation" of the capital within a settlement, it may be prudent to allow a balance to occur as any imbalance especially in the amount of capital in circulation could lead to strained relations between settlements. It may, however, be that the "transformation" of the values into prices of production could have social consequences inside a colony and spread to many generations in particular during an interstellar voyage (Moseley 2000). The products used by the space explorers and the equipment and intellectual property on which they use to produce various "commodities", will see their values transformed into so-called prices of production. In space it will not be possible to reveal the "economic law of motion of the society" as predicted by Marx because the types and levels of labour will not be the same but commodities will still be exchanged based on the labour theory of value.

So what other items of value apart from capital will need to be amassed to ensure a constant flow of money to finance various industrial, scientific and related projects? Will production still be dominated by use-value via consumption or will there be a return to a pseudo-feudalism where the exploiter i.e. finance provider will utilise the value of major discoveries and inventions for their own needs and not those of a settlement? In other words the settlements that need the financial capital to survive will be carrying out the work and the financiers will reap the profits; any surplus value will be ploughed back to shareholders and other backers on Earth. Such an exchange between the two main parties i.e. terrestrial versus extra-terrestrial could create a society where those in the settlements could only survive by virtue of those providing the capital from Earth. To benefit all parties involved, Marx (1975) would argue that any surplus-value should be reinvested in the settlements as long as there was sufficient profit being generated in terms of the financial, economic and intellectual outcomes achieved. Ideally, the more profit that is generated, the more the surplus value will be.

In the colonisation of space, there will come a stage when the capitalist exchange mechanism within a settlement (and possibly in an interstellar spaceship) reaches a critical level and in response there will be a social revolution in terms of the political, financial and economic management of the settlement. Such a revolution could antagonise those on Earth who have invested within a particular settlement and now fear a reduction in profit. However, the evolution of democratic representation within a settlement should ensure better relationships with financial backers on Earth as decisions made are more open and serve the needs of the many rather than a few. Because of the scarcity of resources, there will be limits on those living and working within a settlement on how much they can spend. Any unspent capital will need to be saved or invested wisely to gain financial profit. An excess of capital to spend within a settlement could lead to an influx of entrepreneurs.

7.11 Labour Value and Demand

In "Capital" Volume 1 (1990), Marx has shown that the use of labour is involved in the social and co-operative nature of a business and of society as a whole. Labour produces the goods that society needs so the labour force both as individuals and as a group collective are important. Extra-terrestrially, there will be limitations on how much the labour force will function but the laws of supply and demand will apply. For example, those persons with the essential technological and scientific skills and competencies such as astronauts, astronautical hygienists (Cain 2011) and space physicians will be paid more and given greater power and status whereas those whose function is to provide the basis needs of the settlement as cleaners, cooks etc. will receive lower wages. However, because of the limited resources available to all persons and the necessity of all to co-operate to survive, even the "lower" workers will need to possess a skill set. Nevertheless, this "lower" workforce will still be the bedrock labour supply and therefore socially divorced from the skilled and wealthier astronaut/technocrat class. But because the "lower" labour market will be providing those services necessary for the smooth running of the settlement and hence its survival, the division is not likely to be as devisive as found in terrestrial societies. For the colonisation of space to occur with minimal conflict there will need to be the freedom for individuals to move from one class or division to another.

Marx (1990) explains that the necessity of the distribution of social labour in definite proportions cannot possibly be done away with by a particular form of social production. In space this observation may not apply due to the nature of capitalism and how labour will be structured. It is likely that the labour force will not be part of a central distributive network which plans on how much specified labour will be needed for particular tasks. If the distribution of labour does take such a form that requires central planning then our whole concept of space exploration and the distribution of settlements may take a different form. in particular as the distribution of centrally planned labour may not be efficient or effective to enable the continued survival of a settlement. However, in the initial stages of space colonisation and for the development of settlements to grow there may need to be some central planning so that relevant labour is allocated to specific tasks for example, to ensure that the life support systems are well maintained. As there will be no availability of excess labour, there will be no need for the provision of welfare. Redundant labour in settlements will be ineffective and therefore policies will need to be implemented so there will always be work based on a balance between resource and necessity (Merrifield 2011).

In space there will be a need for "variable capital" that will be used to buy labour and "constant capital" to buy plant, raw materials and equipment. As the populations of the settlements expand, then it is expected that the labour costs will become cheaper. However, because the settlements are likely to be limited in size and have a need to provide all the essentials for survival, it is highly likely that any pool of excess labour will be small and short-lived.

This excess "unskilled" labour would need to be well regulated and controlled to prevent them from causing unrest.

As a drastic measure to ensure that there is stasis in the population, the redundant labour may need to be exported to Earth or to another settlement. This could also include the deportation of any children born to the excess labour force unless productive means could be generated to utilise them. These draconian steps to curb an extra-terrestrial surplus of labour would have profound affects on the freedom and liberty of all sections of society. For those people that were selected to remain in a settlement, the working day would need to be extended to utilise their labour profitably but this need not be at the expense of a reduction in their outside education and activities. To do otherwise, would be against the interests of the colony as a whole and those financing specific enterprises. It is likely that legislation would need to be drafted by those in power to limit working hours and to ensure the mental and physical health of the labour force involved. The labour process could then become socialised with a complex division of labour or what Marx in "Capital"—Volume 2, Chap. 6 (1992a) called the "collective worker".

Terrestrially, the "industrial reserve army" of stagnant workers would ensure the continuing provision of low wages. These workers could train to work in new industries on Earth or be used extra-terrestrially to service others in a settlement or to man a World-ship en route to other worlds. However, for those who decide to work in a settlement, there will be a fluctuating response to rises and falls in the wage rates over time. It is expected economically that any wage rises will not cause an increase in the bargaining power of the labour force. Marx (1990) in Chap. 25 of Volume 1 of "Capital" wrote that "the general movements of wages are exclusively regulated by the expansion and contraction of the industrial reserve army." This is a sweeping statement but it is self explanatory with wide implications in particular for a settlement labour supply if there is limited investment and industry fails to expand at a sufficient rate to generate more wealth. In general, over time it is likely that wages will rise according to the laws of supply and demand. In settlements, there will be a minimum number of workers needed to ensure the provision of the life support services and if there is a demand for more of these services then a larger pool of workers will be needed. The cost of providing the life support services will be by investment by the capitalists, industrialists and financiers. The exploration and settlement of new worlds will generate more profit for all the investors including those private individuals who will have bought shares. Cockell (2013) states that by taking the provision of oxygen as an example, "Any industry that seeks to produce oxygen must invest huge resources in people and infrastructure to extract this resource from rocks or the atmosphere." Space worker wages will rise and not follow the "iron law" of wages where on Earth it has been shown, falsely in many cases that wages don't rise above the lowest necessary for subsistence.

As settlements become well established and thrive, then there will be an associated increase in productivity because the workers available will need to work harder for the benefit of the settlement. This could be offset by an increase in population meaning there will be more free labour over time. However, unregulated births within a settlement could put a tremendous strain on all resources so some form of control will be necessary. The use of new efficient and effective technologies for example, to remove ores from asteroids or to extract water from comets will see fewer workers and therefore an increase in productivity. The labour force via increased productivity will have more money to spend as the falling value of commodities gathers pace. What these commodities will be cannot be predicted but they will need to be marketed to ensure that the public is aware of them by aiming at the correct target audience. Effective marketing of goods at the right price, at the right place and at the right population groups will be critical to ensure a market share in particular as living standards rise within the settlements.

7.12 Labour and Profits

In the many extra-terrestrial settlements that will be established, a drop in profits due to low productivity will not be specific to one section of the economy such as retail or manufacturing nor will it occur at predicted times. But the downturn will be widespread and will relate to a progressive movement of the labour force. This may not affect the highly trained astronauts whose rates of production are expected to be constant and not as progressive as the rest of the labour force. As the productivity of the labour market increases, the need to ensure the efficient and effective working of the machinery and the extraction of raw materials will increase. The investment in plant from Earth or from another successful and flourishing settlement will grow relative to the variable capital used to pay the worker's wages. However, as the productivity increases, the rate of profit will fall. So, because of the unending possibility of expansion in the space environment, there will be a continual need to invest for higher productivity even though it could reduce the profits made. In the short-term, the capitalists will benefit from high productivity especially if the profits can be fed into the development of one or more settlements but in the long-term this may not be the case unless competition can be curtailed and the exploration of space expands. At this stage of space conquest, it cannot be predicted how Marx's theory of profit will be applied or even if it will apply extra-terrestrially.

Extra profits from prices could be found by price fixing at a level that is lower than the social value. If each settlement is governed as individual social states, it may be difficult to interfere with the price mechanism and the capitalists may therefore need to seek other means for generating extra profits. The capitalists, industrialists and space explorers will need to utilise any new technologies for example, smart robotics to stop themselves from being undercut and subsequently driven out of business. If the settlements are in competition with each other and the product prices fall as the market winners become more commonplace and others produce them cheaply, it is only then that there will be the potential to be pushed out of the market. As yet, it is not possible to predict whether the initial advantages gained by being the market trailblazer for one or more commodities will have the same impact extra-terrestrially but if they do it will be because of effective commodity marketing and competition between settlements for intellectual property. If there were no market competitors, then it could be argued that there was no need for labour productivity to increase resulting in economic stasis. The social and political structure of a settlement, the labour force make-up together with the worker skills and abilities and the scientific functions of a colony will influence the level and degree of competition needed to successfully flourish as a viable colony. The law of value by itself may not be sufficient to increase the labour productivity but if it is necessary, then good communications that cover a wide space area and faster means to traverse space will be important for promoting new commodities.

Living and working in space must provide a suitable environment for all capitalists and financiers to operate profitably for the betterment of all the space settlements and to avoid the need for increasing the amount of surplus-value for each worker and each competitor. Without the application of such economic principles as hypothesised by Marx, it may be impossible for each settlement to generate sufficient wealth to survive unless there was an influx of investment from Earth. In "Capital"—Volume 3, Chap. 15, Marx (1992b) states that ".....every such new method of production cheapens the commodities. Hence the capitalist sells them originally above their prices of production, or perhaps, above their value. He pockets the difference……". In such situations, the need for money or an alternative means of barter will need to exist either as a physical object or by the use of cyberspace transfer. The laws of supply and demand and the need of profit as outlined by Marx will probably manifest themselves in unpredictable ways but there will be a developing need to manipulate the environment for profit. This manipulation may eventually cause a drop in the rate of profit as production becomes greater leading to increases in labour productivity and the potential for an economic crisis.

7.13 Labour Rights

Because the rate of surplus-value has risen, the position of labour in relation to other workers may worsen. As the level and degree of production expands, having a surplus worker population will strengthen the power of the employers and those sponsoring the scientific and industrial ventures. This could lead to the exploitation of the surplus labour force to increase production and thereby extend the rates of profit generated for investors. Marx was producing his economic theories when labour was defined as mainly uneducated manual workers. But by its very nature, those working and living extra-terrestrially will need to be well educated and be expected to understand the laws of economics as they apply to their circumstances. They will have to understand how capitalism works in society to produce the maximum profits and possibly at the expense of the working population. They will want the instigation of laws to regulate the worst excesses of capitalism (Sweeney 1968). This will lead to negotiations between the various interested parties either through "Unions" (if they still exist as bartering bodies) or their equivalent. If there is obstinacy on all sides during negotiations for higher wage settlements and conditions of work then there could be a crisis.

Because of the hostile environments in which all space persons live and work, the outcomes of negotiations that are suitable for all sides would need to be quickly decided. However, if there was a failure to reach a speedy settlement then the results could spread to workers in other colonies who would fear for their own jobs.

Critics of Marx such as Hayek (1960) and Friedman (1957) have argued that rising labour productivity will not lower the profits generated; the rate of profit will not fall. The capitalist will therefore make a profit on his original investment. The means of production will have been bought at the original values and not at the lower labour time it would now cost to produce them. In the highly volatile environment of "space capitalism", it is highly likely that the plant, equipment and human resources will be utilised within a short period of time and it will be the capitalists' continuous response to the changing environment where profits will be made rather than in classical economics where an investment would expect to see profits generated over many years.

7.14 Extra-Terrestrial Economic Crises

In "Capital"—Volume 3, Marx (1992b) explains how a reduction in the money supply and a fall in profits could lead to an economic crisis. Such crises could occur extra-terrestrially. Inside settlements there will only be a demand for a commodity if the demand already exists and there is purchasing power to buy the goods. However, who will set the trends for the need to buy specific goods and who will pay for the marketing in particular in relation to the manufacture of high technological, engineering and hard ware goods that will be required to survive. The lack of supply of certain goods produced by a settlement and sold terrestrially could lead to significant concerns but they are likely to be limited in scale because each independent settlement should be able to "fine tune" their economies to produce only those goods in sufficient quantities that can be sold terrestrially. Without such fine-tuning, an economic crisis could occur with consequences both terrestrially and extra-terrestrially.

How will an economic crisis arise in one or more space settlements and how will they relate to the Marxian concepts of crises as they occur on Earth (Harman 1981)? If workers in the manufacturing industries have a cut in their wages to increase the rate of surplus value, they will have little money to spend on consumer goods. Terrestrially, this may be a minor setback unless there are no wage increases over a prolonged period, but in space the need to buy and have access to certain consumer goods (e.g. communication equipment) that are not produced extra-terrestrially could be critical. If capitalists producing these goods react to a lack of demand, they may decide to cut back orders for new plant and equipment leading to a fall in demand for the products and a need to lay off workers. This will have a knock-on effect and may require a Keynsian economic approach to address the issue for example, by governments printing more money and by carrying out more public works (Keynes 1936). The knock-on effects on Earth would have implications for the management of extra-terrestrial colonies unless they had tools available to counteract the crisis.

Both the terrestrial and extra-terrestrial economies would need to run parallel to ensure that when an economic crisis occurred in one it would not impinge on the other and the economy of one could be used to "kick start" the other. However, it has to be accepted that because of the economic nature of commodities, any economic crises are most likely to occur repeatedly both terrestrially and extra-terrestrially. In general, if a commodity is sold, the money is used to buy other commodities. But what if the seller wishes to keep the money made by selling commodities and not invest the profits? This will occur if the rate of profit from selling a commodity is too low for it to be worth investing (Bakir and Cambell 2006). If this was to happen amongst many buyers and sellers, then there would be a shortage of money and a depression or slump could occur. Any number of events such as a banking crisis could tip the balance between selling and investing the profits. As Marx stated in "Capital", Volume 2, Chap. 21, (1992a), "a balance is itself an accident owing to the spontaneous nature of production". Would the unbalancing of the economic system have similar properties whether it occurred either terrestrially or extra-terrestrially and would stasis be more easily controllable in the settlements? If the labour market working inside a settlement or employed on a starship became more productive, and this may be difficult to attain in the space working environment, then the rate of profit could fall even with safeguards guaranteed by a capitalist democratic regime.

What does Marx perceive to be the definition of an economic crisis? Like Perelman (1987) he sees it as a scarcity of labour and finance that prevents the investment of money into an economy. What scenarios could occur during the human settlement of the Moon and the planets that could create an economic crisis similar to that on Earth? As examples, if there was a sudden rise in the value of the energy required by a settlement, Marxian and capitalist economics would indicate that there may be a cost increase in the import of raw materials to offset this. A significant shortage of scientists and specialist engineers needed to service and maintain vital equipment could also cause a crisis. However, because of the need for colonies to be self-sufficient, a banking crisis or stock market crash on Earth would take time to cause major affects on the economies of settlements. Extra-terrestrially, there would be time to monitor and mitigate the affects of such a crisis. But those settlements established on the Moon and Mars and therefore in the vicinity of terrestrial influence would be at a greater risk from the affects of an economic crisis.

A single large settlement that had its own currency and banking system would be especially hit by an economic crisis in particular if there was a high volume of trading with Earth. In this situation there would need to be mass migration to other settlements to reduce the affects of the crisis and to ensure the sustainability of the settlement. Lessons may be learned from Marx in such situations for example, the banks could provide additional finance and related resources to stimulate the economy of the settlement but this could exacerbate the crisis if it was not well managed.

Because of human greed, a crisis could occur in a settlement caused by the "hoarding" of vast amounts of cash and commodities by individuals or by the Government. If there was panic within a settlement population then "hoarding" could spread leading to hardship and possibly the destruction of the settlement unless outside finance was invested to stimulate the economy and cause investment. It could be argued that when settlements are widespread throughout near and interstellar space, the economic crises may be on a grandeur scale and follow economic patterns that are not discussed by Marx in his works. The concept of Marx as outlined in "Capital"—Volume 1, Chap. 24 (1990) that the "limits of production are set by the profit of the capitalist and in no way by the needs of the producer" may not apply. However, with sophisticated economic modelling, it may be possible to predict when a crisis may occur and take mitigation actions to prevent it occurring. But the reverse may happen; the possibility of a crisis occurring may lead entrepreneurs to take unnecessary risks.

Marx outlines in Volume 3 of "Capital" (1992b) that there is likely to be a depreciation or devaluation of capital as a crisis develops and this may be true of such a crisis in one or more planetary settlements. Vast amounts of capital will be taken out of the economy and businesses and services that have thrived may "go out of business". There may be an implosion of confidence amongst the extra-terrestrial settlements leading to a reduction in exploratory expeditions and the development of new technology. Eventually, all economic crises will pass leaving behind tales of human misery. Such tales may form the background for new space myths and the introduction of different art forms to enable the human spirit to cope. The length of time of a crisis and the time taken to be eradicated will vary from settlement to settlement but constant capital will eventually reach equilibrium and there will be a depreciation of the value of goods. The cost of capital will also be reduced and the rate of profit will eventually recover.

7.15 Economic Crises and Labour Welfare

This profitable employment of capital will have affects on the labour force who may migrate to those settlements where better initiatives and bonuses in terms of living conditions may occur. So even though a crisis may be receding, there could still be long-term repercussions on individuals as the labour market fluctuates. The extra-terrestrial labour market is likely to be vast and cover settlements over a large area of space The skilled scientists and engineers will have a much greater opportunity for bargaining and therefore will be able to win more advantageous benefits in terms of status and health. The provision of excellent health facilities to specific groups of privileged workers will be a major factor that can be used by employers to attract the best workers.

It is highly likely that because of the nature of the labour market and the costs involved in providing effective health care to all workers (Gordon 1972), that there will develop a two tier system of health care with those unskilled workers receiving poorer health care than skilled workers (Cain 2014). Such a division in labour will cause division throughout the settlements as the highly skilled workers receive the better benefits and incentives. This division could also affect settlement governance and the liberty of the population as a whole who may see a diminution of their democratic rights. Many sectors of a settlement population could be disenfranchised in particular those whose skill set is not sufficient to contribute to the future development of the settlement. Universal suffrage may no longer apply except in isolated settlements where stasis in society is essential for the survival of all. If settlements do evolve with a limit on democratic freedoms, then a system of governance based on Marxist principles may well become the political system of choice by those that govern and not that based on the principles of capitalism as proposed by Hayek (1960) and Friedman (1957).

An economic crisis throughout the humanised sections of the Universe could exacerbate the destruction of those social systems that adopted full democratic beliefs though the implementation and application of such beliefs may function differently in a two tier system of labour. The economic crises will occur in those periods when the capitalist system appears to be offering the best means to ensure settlement survival and development. Unless the system is re-organised and re-shaped to quickly restore the rates of profit to a level at which effective investment will occur then those disadvantaged members of a settlement society will suffer. The stronger settlements in terms of those receiving regular inward investment, those providing stable governance and those promoting democratic rights will be in a better position to survive and flourish in an economic crisis. The weaker settlements who will be most affected by an economic crisis will see stricter controls on their freedoms and the possibility of a Fascist regime to restore their "national" esteem. This is to be avoided by the implementation of effective measures to reduce the affects of the crisis. The pooling of human, economic and physical resources between neighbouring settlements could be one such measure and lead to a degree of centralisation as the smaller colonies become incorporated into the larger ones or the more efficient colonies "take over" those less efficient economically.

7.16 Marxism V Capitalism—An Extra-Terrestrial Battleground

Time, experience, an evolving research and task framework and other factors some known and some unknown will decide whether the economic pattern of the early space explorers will follow a capitalist system, a Marxist system or some other economic paradigm. Whatever economic theory is consciously or unconsciously chosen as the best suited for extra-terrestrial survival will be driven by history and necessity; but at some stage it will be held up to scrutiny by future economists and historians. Such intellectual scrutiny will be important in judging and selecting the most adaptable political and economic system for one or more settlements and it may incorporate social factors, work patterns and the utilising of the scientific and technological advances for the benefit of all. How will the concepts of social working and communism apply to extraterrestrial settlements and to travellers on interstellar space ships? Will capitalist theories and ideas as postulated by the neo-liberals interact with Marxian concepts to create a new economic system adapted to life in space? These and other questions await an answer.

The use of capitalist principals to regulate the world's financial, economic and social systems has been successful in bringing prosperity to many parts of the world. However since the banking crisis in 2008 where the developed world's financial system nearly collapsed there has been a questioning as to whether capitalism linked with democracy is the best economic system for social progress. The success of China economically now indicates that this model may not be the best to apply to the majority of the world's population; the application of Marxist socialist doctrines together with "raw capitalism" may be the best system for progress from a practical perspective. This model may also be the most successful for extra-terrestrial governance where a strong central government may be needed for survival.

New political thinkers such as Mulcan (2009), Gordon (1972) and Przeworski and Limongi (1993) have been influenced by a blend of Marxist and classical philosophy. They have adapted neoclassical economics to incorporate Marxian concepts. Their ideas have been incorporated into how governments plan and finance large projects such as the building of space ports. Both Marx and Engels (2011) understood that capitalism was only one staging post in a succession of economic systems that would eventually lead to communism. In the first instance it is highly likely that extra-terrestrial governance will incorporate a form of capitalism together with socialist ideals but there may be a "sea-change" in economic thinking to adapt to the harsh environment of planetary colonisation. Those settlements that take risks in developing their social needs and exploit a planet for profit will be at the forefront of change leading to competition and challenge on an unimaginable scale not only economically but also in the arts and sciences.

The exploration of space and the understanding of Mans' future role in this goal will at first be detached from any particular mode of economic system. For terrestrial policy-makers who will need to interpret events as they occur in the space settlements, there is likely to be an "elastic understanding" of capitalism and a subsequent entrenchment of the market forces using financial tools developed on Earth. Such entrenchment will not succeed unless national strategists reach a "Nash equilibrium" i.e. no player has anything to gain by changing their own strategy (Nash 1951). Communications between all parties in particular the colonists themselves will require a "gentler" kind of capitalism to understand events and the need for sustainable change. As seen from Earth, the extra-terrestrial societies may appear stable but closer analysis may reveal the opposite (Kaye 1988).

7.17 Conclusion

In the struggle to survive in space, certain businesses will thrive as competition between settlements takes affect and capital is accumulated by wealthy, entrepreneurial individuals. Periods of boom and bust will need to be avoided, but this may not be possible because of the links to the capitalist system of economics and finance. However, the colonisation of space will bring unpredictable consequences both for the economic systems utilised and in the systems of governance implemented. In Marx's "Theories of Surplus Value" (1975) he indicates that the "development of the richness of human nature is an end in itself". This is a fine sentiment (Burke et al. 2011) and could be applied to the settlements of the Moon, Mars and other worlds. But this is not an endorsement of Marx and his principles as outlined in "Capital" Marxist theory provides just one interpretation of the use of capital and the development of the labour markets. It is right that in this early stage of Man's exploration of space that Marxism should be analysed critically so that his theories, so tragically and forceably implemented on Earth are not repeated as Man colonises other worlds (Callinicos 1983).

References

Bakir, E., & Cambell, A. (2006). The effect of neoliberalism on the fall of the rate of profit in business cycles. *Review of Radical Political Economics, 38*, 365–373.

Bellofiore, R. (1989). A monetary labour theory of value. *Review of Radical Political Economics, 21*, 1–25.

Burke, J. P., Crocker, L., & Legters, L. H. (2011). *Marxism and the good society.* Cambridge: Cambridge University Press.

Cain, J. R. (2011). Astronautical hygiene—a new discipline to protect the health of astronauts working in space. *JBIS, 64*, 179–185.

Cain, J. R. (2014). Astronaut health—planetary exploration and the limitations on freedom. In C. S. Cockell (Ed.), *The meaning of liberty beyond earth.* New York: Springer.

Callinicos, A. (1983). *The revolutionary ideas of Marx.* London: Bookmarks Publications.

Cockell, C. S. (2008). An Essay on extraterrestrial liberty. *JBIS, 61*, 255–275.

Cockell, C. S. (2009). Liberty and the limits to the extraterrestrial state. *JBIS, 62*, 139–157.

Cockell, C. S. (2013). *Extraterrestrial liberty: An enquiry into the nature and causes of tyrannical government beyond the earth.* Edinburgh: Shoving Leopard.

Daily, G. C., & Ehrlich, P. R. (1992). Population, sustainability and earth's carrying capacity. *BioScience*, 42, 761–771.

Friedman, M. (1957). *A theory of the consumption function.* Princeton, NJ: Princeton University Press.

Gordon, D. M. (1972). *Theories of poverty and underemployment, orthodox, radical and dual labour market perspectives.* Mass: Lexington Books.

Harman, C. (1981). Marx's theory of crisis and its critics. *International Socialism, 2*(11), 30–71.

Hayek, F. (1960). *The constitution of liberty.* Chicago: University of Chicago Press.

Jameson, F. (1984). Postmodernism, or the cultural logic of late capitalism. *New Left Review, 146*, 53–92.

Kaye, F. B. (Ed.). (1988). *The fable of the bees.* Indianapolis: Liberty Press.

Keynes, J. M. (1936). *The general theory of employment, interest and money*. London: Macmillan and Co Ltd.

Leysens, F., Driesen, K., & Wouters, K. (2010). On the organisation of world ships and other gigascale interstellar space exploration projects. *JBIS, 65*, 134–139.

Marx, K. (1975). *Theories of surplus value—part 3*. Moscow: Progress Publishers.

Marx, K. (1990). *Capital: Critique of political economy—Volume 1*. London: Penguin Classics.

Marx, K. (1992a, May). *Capital: Critique of political economy—Volume 2*. London: Penguin Classics.

Marx, K. (1992b, August). *Capital: Critique of political economy—Volume 3*. London: Penguin Classics.

Marx, K., & Engels, F. (2011). *The Communist Manifesto -1848*. New York: Penguin.

Merrifield, A. (2011). The right to the city and beyond: Notes on a Lefebvrian reconceptualisation. *City Analysis of Urban Trends Culture, Theory, Policy, Action, 15*(3–4), 473–481.

Merrifield, A. (2013). The urban question under planetary urbanisation. *International Journal of Urban and Regional Research, 37*, 909–922.

Moseley, F. (1997). The development of Marx's theory of the distribution of surplus value. In F. Moseley & M. Campbell (Eds.), *New Investigations of Marx's Method.* Atlantic Highlands, NJ: Humanities Press.

Moseley, F. (2000). The new solution to the transformation problem. A sympathetic critique. *Review of Radical Politics, 32*, 282–316.

Mulcan, G. (2009). *The art of public strategy*. Oxford: Oxford University Press.

Nash, J. (1951). Non-cooperative games. *The Annals of Mathematics, 54*(2), 286–295.

Perelman, M. (1987). *Marx's crises theory. Scarcity, labour and finance*. USA: Greenwood Publishing Group.

Perelman, M. (1993). The quantitative side of Marx's value theory. *Rethinking Marxism, 6*, 82–95.

Perelman, M. (2010). Constant capital and the crisis in contemporary capitalism: Echoes from the late nineteenth century. *The Journal of Economic Analysis, 1*, 34–41.

Przeworski, A., & Limongi, F. (1993). Political regimes and economic growth. *Journal of Economic Perspectives, 7*, 51–69.

Sawyer, S. (2012). The future and Stapleton's visions. *JBIS, 65*, 25–29.

Sraffa, P. (Ed.). (1951). *On the principles of political economy and taxation: In works and correspondence of David Ricardo* (Vol. 1). Cambridge: Cambridge University Press.

Sweeney, P. (1968). *The theory of capitalist development.* London: London Publications.

Chapter 8
Space Polities: Self-governance Lessons from Virtual Worlds

John Carter McKnight

Abstract This work argues that the virtual worlds of the mid-2000s can be read as analog environments for libertarian space settlements, as their designers shared a common ideology, and were often explicitly inspired by, advocates of such settlements. These worlds confronted questions of creating new cultures in an environment hoped to be outside the terrestrial legal and governmental system, heavily shaped by its enabling technologies, and capable of experimenting with social systems incapable of enactment on earth. Unfortunately, virtual worlds failed to produce libertarian paradises, for reasons with profound implications for space industry and settlement. Designers implemented engineering systems at odds with their social values, failed to enable social, rather than technological, solutions to emergent social problems, and failed to understand the needs and values of residents of these new socio-technical environments. Residents, on the other hand, outright rejected the libertarian, anarcho-capitalist communities created for them by designers, also rejecting citizenship-based state models in favor of corporate managerialism. Since the governmental and social models users rejected are those most commonly put forward by advocates of extraterrestrial governance systems, it may be time to explore new models capable of inspiring developers and sustaining future users alike.

Keywords Governance · Ideology · Social science modeling · Virtual worlds · Space settlement

This paper is built in part upon material presented in October 2014 at the One Hundred Year Starship Public Symposium as "Beyond Heinlein: Facing the Space-Libertarian Shortage." Jemison, M., Batt, J., and Almon, A., eds. *Proceedings of the 100 Year Starship 2014 Public Symposium.* The Dorothy Jemison Foundation for Excellence. (forthcoming)

J.C. McKnight (✉)
Bowland North, Lancaster University, Bailrigg, Lancaster LA1 4YN, UK
e-mail: j.mcknight@lancaster.ac.uk

C.S. Cockell (ed.), *Human Governance Beyond Earth*,
Space and Society, DOI 10.1007/978-3-319-18063-2_8

8.1 Introduction

From the fictions of Jack Williamson in the 1930s through Robert Heinlein's (1966) work to a range of late-1990s nonfictional commentary, much speculation around space settlements has engaged with the notion of liberty, particularly in positing extraterrestrial communities as a haven or laboratory for systems of greater personal liberty than currently seen on Earth. While some have pointed out that the fragility of space infrastructure, its likely high cost, and the lack of adequate indigenous resources including air may predispose space vessels to tightly-controlled dictatorships, the anarcho-capitalist or libertarian community has remained an common model in fictional and non-fictional speculation alike. Meanwhile, from the 1990s, as private enterprise in space was being advocated by activists of a distinct libertarian bent, the internet was similarly constructed in techno-libertarian rhetoric as a frontier apart from control by nations of the Earth, in which people could be free from the ideological limitations of national governance and the limitations of identities inscribed on their bodies. Thus, when a generation of virtual worlds, some explicitly designed as havens for libertarian experiments came on line in 2003 and subsequent years, it became possible to examine them as experimental testbeds for the political views and practices advocated for space settlements as well as for the internet.

A 5-year study of a range of communities and organizations in the most popular gaming virtual world, World of Warcraft, and the most popular non-game world, Second Life, found a consistent pattern of outright rejection of libertarian and collective self governance models of social organization in favor of hierarchical and managerial models in which the great majority of people delegated authority to a supreme leader. Additionally, rather than debating and adopting rules, in many cases the chosen means of dispute prevention and resolution was via software, in a delegation of authority not even to other human beings but to machine algorithms.

Virtual worlds were far from a complete analog for space settlements: participation was voluntary and rarely 24/7, with negligible barriers to exit from particular communities or from the platforms. The parallel to be drawn is not to near-term space communities but to those for which the libertarian model has been advocated: large permanent settlements within a network of diverse alternative habitats. The lessons to be drawn from the virtual experience are twofold: there is a shortage of would-be libertarians, and many people drawn to innovative, highly technological environments tend to prefer to solve social problems by technological, rather than quasi-governmental, means.

This paper briefly sketches the history of space-libertarian thought in speculative fiction and non-fiction and its relationship to notions of governance on the internet. From there, it summarizes 5 years of fieldwork on the governance of virtual worlds, explaining witnessed outcomes on the basis of trends in a broader Western middle-class technologically sophisticated culture. It suggests that the failure of uptake of the most commonly advocated ideological system for space settlement provides an opportunity to examine the question of extraterrestrial governance anew, and to create or examine other potential models for their applicability to questions of human social organization in space.

8.2 Space as a New Frontier of Liberty: An Overview of Speculation

The modern linkage between space settlement and libertarian movements may have begun with *Birth Of A New Republic*, co-authored by Jack Williamson. (Williamson and Breuer 1981, [1931]) The novel describes a movement inspired by the American Revolution to overthrow a repressive corporate state on the Moon. Recasting the American Revolution in space is something of a natural what-if scenario, analogizing to the distance and hardships in traveling to and sustaining settlement for Europeans in North America, and building on a tradition of political thought dating back to de Toqueville (1835, 1840 [2003]) and Frederick Jackson Turner (1893, in Faragher, ed. (1994)), of explaining the greater political freedom of the English-descended settlements in North America through the effects of distance from the metropolis and the personal resilience required to inhabit a challenging environment.

The theme was most famously employed in Robert Heinlein's (1966) *The Moon is a Harsh Mistress*, perhaps the most famous and influential work on liberty and extra-terrestrial settlement. The novel envisions the Moon as a penal colony along the lines of Australia, in which the convicts and their descendants revolt, issuing a declaration of independence on July 4 and creating a congress. However, Heinlein presents an indigenous culture based on libertarian principles: in the absence of conventional state institutions (as the Warden's Authority only maintains life support and infrastructure, leaving the inhabitants to their own devices), the inhabitants have developed the minimum necessary mores and customary procedures for social functioning. As they are all prisoners, convicted or de facto from musculo-skeletal weakness of Lunar gravity, freedom from authority is prized. Respected community members may be called on at any time to adjudicate disputes, with the disputants offering payment, and judgment, including capital punishment, executed on the spot by the assembled crowd. Reputation systems take the place of regulatory schemes and custom takes the place of law, a common arrangement in relatively isolated communities in which neighbors can count on decades of repeated interactions with each other. (Heinlein 1966).

Heinlein's work marks a distinction between concepts of liberty applied to space settlements: between anti-colonialist or anti-socialist visions for which the early United States serves as a model, and anti-statist libertarian imaginings. For the former, John Locke (1689) supplies a theoretical foundation; for the latter, Richard Nozick (1974), or, on the left, Murray Bookchin (1991). Despite the rise of private commercial spaceflight and the articulation of an anti-statist ideology for internet governance in the 1990s, and significant cultural and generational divisions between advocates of the two ideological perspectives, statists motivated by American nationalism and libertarians inspired by a vision of a break from state power have tended to cooperate within the space-advocacy movement. For both, analogies to the Euro-American conquest of the American West are commonplace.

That the analogy to the frontier of the American West has been a primary conceptual framing is beyond doubt, from President John F. Kennedy's repeated invoking of the term though *Star Trek*'s "Space, the final frontier," in the years (1966–1969) leading up to the moon landing. A key policy document from 1986 is entitled *Pioneering the Space Frontier*, and elaborates extensively on the analogy. (National Commission on Space 1986) While the meaning and value of the frontier as historical legacy and cultural metaphor for space exploration has been highly contested, (see, e.g., McCurdy (1997) and Limerick (2000)) the debate over the cultural meaning of the Western frontier and its consequences for space exploration is beyond the scope of this work.

There has been, however, an explicit argument for extraterrestrial settlement directly informed by a sophisticated reading of Turner. Robert Zubrin, founder of the space exploration advocacy group The Mars Society, argued in a 1994 essay for Mars as space of political liberty in a frontier, rather than techno-libertarian, vein. The essay begins with an evocation of Turner delivering his 1893 paper. Zubrin argues that Turner's conclusion that the frontier had closed may have been premature then, but was proving true a century later, and that without a new frontier, "progressive humanistic culture" is fading. The key to a frontier is that it be remote enough to allow for the free development of a new society, and that everywhere on Earth, "the cops are too close." Mars, however, has both the cultural distance and the potential wealth to create a new society driven by an engine which destroys aristocracy and institutional stagnation, and promotes democracy, diversity, and individual dignity (reprinted in Zubrin 1996).

In sharp contrast to Zubrin's American-exceptionalist view of Mars as a site of reinvigorated national culture, an argument for remote spaces of communal scientific research as a space free of the constraints of corporate capitalism—similar in some respects to the vision of Williamson and Breuer (1931) has been theme of much of novelist Kim Stanley Robinson's work, particularly *Antarctica* (1997) and his Mars trilogy (1992, 1993, 1996). Robinson argues for scientific communities as a model for a socialist utopia—quite literally the Republic of Science (Polyani 1967).

On the libertarian side, O'Neill's 1976 book *The High Frontier* is a foundational work in space settlement advocacy. Drawing on a model apparently coincidentally similar to Nozick (1974) work, he advocates a system of great numbers of space habitats in the Earth-Moon system, each a platform for social experimentation, with free travel among them, in effect creating a free market in governance systems. O'Neill's work was crucial in inspiring the American space advocacy movement of the next several decades, including such organizations as the Space Studies Institute; the L5 Society, which later merged into the National Space Society; and the Space Frontier Foundation. The latter organization in particular rose to prominence in the late 1990s advocating a capitalist, libertarian approach to spaceflight at the time of the dotcom boom, as several entrepreneurs who made their fortunes in information technologies and applications turned their attention to spaceflight. (e.g., Elon Musk of SpaceX, Jeff Bezos of Blue Origin, and John Carmack of Armadillo Aerospace) Many advocates of both space and internet technologies in the era shared a common ideology, (Barbrook and Cameron 2007) one viewing both technologies as critical enablers of personal liberty.

8.3 The Internet: Electronic Frontier of Liberty

The 1990s saw the application of libertarian beliefs to the internet, and the beginning of a techno-libertarian movement drawing in part from American Progressive technocratic ideologies, the counter-culture movement of the late 1960s, the belief that social problems could be addressed via software engineering, free-market absolutism, and a distrust of state institutions, as Turner (2008) documents. Again the frontier metaphor is employed, but in an anti-statist fashion, rather than the imperialist, Manifest Destiny, reading given it by Zubrin. The phrase "the electronic frontier" was coined by John Perry Barlow, a Montana cattle rancher (and Grateful Dead lyricist). Barlow took the popular notion of "cyberspace," from William Gibson's seminal novel Neuromancer (1984), which depicted a global information network perceived as physical space, and then described that space in Turnerian terms: "Indeed, one of the aspects of the Electronic Frontier that I have always found most appealing—and the reason Mitch Kapor and I used that phrase in naming our foundation—is the degree to which it resembles the 19th Century American West in its natural preference for social devices which emerge from its conditions rather than those which are imposed from the outside." (in Ludlow, ed. 1996) In his 1996 "Cyberspace Independence Declaration" Barlow calls on the governments of the world to "leave us alone," as cyberspace, "the new home of Mind," is the abode of virtual selves immune to state sovereignty, "that all may enter without privilege or prejudice accorded by race, economic power, military force, or station of birth."

Turner (2008) makes a strong case for the direct ideological and social linkages between San Francisco Bay Area counterculture and the culture of 1990s internet "pioneers," one which, via the role of Electronic Frontier Foundation co-founder and investor Mitch Kapor, continues into Linden Lab (LL), corporate creators and owners of SL, as Kapor was chairman of LL's Board of Directors (and delivered a speech in SL in 2008 explicitly declaring the SL "frontier" closed: Kapor 2008), attributing an ideological pedigree to the financiers and designers of SL is insufficient to account for its general perception. For that, one needs to examine the "cyberspace" concept and the work of its key progenitors, William Gibson, Neal Stephenson and Sherry Turkle. Gibson coined the term in his short story "Burning Chrome," (Gibson 1982) and popularized it in his influential novel *Neuromancer* (Gibson 1984), which established not only a range of tropes of the cyberpunk literary genre, but arguably shaped the later development of the World Wide Web.

Neuromancer presents a global computer network perceived by users as an all-encompassing sensory experience of data presented spatially, a "graphical representation of data" in "constellations," and in an often-quoted phrase, a "consensual hallucination." (Gibson 1984, p. 69) Stephenson elaborated on the spatiality of data in his novel *Snow Crash*, (Stephenson 1992), in which networked data was presented in a city-like environment, complete with designer avatars and virtual nightclubs. SL founder and initial CEO Philip Rosedale has described *Snow Crash* as a key inspiration for SL. (Dubner 2007; Sydel 2010) Similarly, Vinge's (1981, 2001) "True Names" established tropes of avatar pseudonymity, identity experimentation and the spatiality of data, and was particularly influential among adherents of the California Ideology: a 2001 reprint volume (with the subtitle "and the opening of the cyberspace frontier") attested to its influence with essays from Marvin Minsky, Danny Hillis, Richard Stallman, and Morningstar and Farmer, creators of the first social virtual world and authors of the first exploration of emergent culture in such a space (Morningstar and Farmer 1991).

In popular nonfiction, Sherry Turkle's (1997) widely-read *Life on the Screen*, prefiguring Castronova's work a decade later, sought to police a boundary between the "real," which for her meant physical, and the "virtual," or actions mediated by a computer. That this distinction existed in the first place is deeply strange: previous electronic communications platforms such as the telephone did not generate a similar notion of a distinct, bounded spatiality of use. The only coherent explanation of this boundary-drawing lies in the prefiguring of the technology as spatialized, via cyberpunk.

Taken together, by 2003 certain notions were well-established in popular, engineering, and academic culture: the spatiality of data presentation, alternative online identity exploration, and a semi-porous boundary between separate realms of the physical and digital. The first generation of virtual-worlds platforms (following after the widely-documented MUDs of the 1990s, e.g. Cherny 1999; Dibbell 1999), including Habitat, documented by Morningstar and Farmer (1991), and Everquest, by Castronova (2001), established additional concepts as central to the genre: emergent behavior, particularly economic, and a sense of the environments as a distinct place where one lived, even preferentially to the physical world. These factors contribute to an explanation of the content of imaginings of virtual worlds; the widespread belief that avatarization and 3D environments were the future of the internet (e.g., Gartner 2007) can best be attributed to 20 years of spatialized, avatarized environments being presented in popular fiction as what the future would look like: strangely, corporate hype was a manifestation of science fiction tropes.

In addition to the Electronic Frontier Foundation, the 1990s saw the emergence of a similarly-named organization, the Space Frontier Foundation, which was a principal advocate of libertarian-capitalist space ventures. Both groups attracted Silicon Valley engineers, investors, and advocates with a common set of political views and values sometimes called the Californian Ideology, or techno-libertarianism. (Barbrook and Cameron 2007). Notably, after the landing of the first flight of a private spacecraft, SpaceshipOne, its pilot unfurled a banner reading "SpaceshipOne, Government Zero" (Foust 2010). Techno-libertarianism was a common thread in envisioning social relations in outer space as much as cyberspace. It is this convergence of software and aerospace engineering advocacy within a common ideology that suggested virtual worlds as a test case for the practical application of a politics long advocated for space settlement.

8.4 From Free Land to the Sleeping Policeman: Engineering Non-human Enforcers and Systemic Constraints

Linda Billings (2007) observes that for both libertarians and American exceptionalists, the concept of the "frontier" has been an integral part of the affective and rational case for space exploration for generations. As noted above, a similar rhetorical appeal was made for the "space" of the internet in its decade of explosive growth. Arguments for space and the internet as a particularly suitable environment for libertarian governance tended to draw heavily on the example of the American West, suggesting and exploring in greater or lesser detail the nature of those environments and how their commonalities allowed for a useful analogy. Where the previous sections of this paper have shown that there *was* a belief that the American West, space, and the internet were envisioned as analogous spaces of potential liberty, this section will explore the why, mapping out the extent to which the analogy between space and the internet for explorations of liberty is tenable, and pointing towards areas in which the (negative) lessons learned in this study may point up shortcomings for the argument for liberty in space.

Many of the arguments for extraterrestrial liberty (e.g., Zubrin 1994, and to a limited degree Heinlein (1966) and Robinson (1996) rely on physical distance from the metropolis to the frontier, on the analogy of the difficulty of Britain in controlling its North American colonies, or the emergent United States in controlling the lands to its west. A more sophisticated reasoning (including a closer reading of Turner (1893) than is often undertaken) is based on the frontier as a *different kind* of space from the metropolis. Here Foucault's (1967) concept of "heterotopia" is crucial to linking certain sorts of colonial settlements with space and virtual worlds, seeing liberty as a prospect when the lack of ability of the metropolis to closely control the frontier—through distance, inattention, or lack of appropriate technologies of governance—combine with the development of local knowledge of the particularities of the environment, including ecological, infrastructural, and cultural factors.

This perspective leads to counter-arguments to the libertarian hypothesis in the case of space, arguments useful in evaluating the analogy of virtual worlds. Heinlein himself touched on critical aspects in his later works, including *The Cat Who Walks Through Walls* (1985): space stations in particular are "inherently political technologies" in Winner's (1980) sense, in that the threat of annihilation from failure of the infrastructure and the likely scarcity of resources makes it likely, if not inevitable, that a strict control regime will govern (an extreme case being Godwin (1954), a short story in which the narrow margin of resources on a spacecraft necessitated the execution of a stowaway). Much of Cockell (2013) is an expansion upon this argument. Where simplistic applications of the American frontier analogy tend to overlook the impact of technological infrastructure on emergent governance systems (compounding a mistake of Turner's (1898), in which he lumps under the term "free land" a complex sociotechnical assemblage of the United States Army, the Conestoga wagon, Jeffersonian traditions of citizen-farmer governance, and a critical misunderstanding of the ecology of the Great Plains, among other factors), the Godwin-Heinlein-Cockell lines of argument tend towards technological determinism and a corresponding under-estimation of the continuity of culture in "new" spaces. This perspective is common to many arguments for and against space as a site of emergent liberty: whether the key factor is seen as distance or delicate infrastructure, two factors tend to be overlooked. First is the extent to which infrastructure enacts previously constituted ideologies; second, that there is no shedding of culture upon immigration but rather a complex interaction of old culture and new environment, which is equally capable, as Foucault (1967) describes, of generating stultifying tyranny as generative liberty, depending on the specifics of the elements in play.

"The Cold Equations" (Godwin (1954) [2003]) epitomizes the former problem. In the story, a stowaway on a spaceship delivering relief supplies to a colony suffering from a plague must be killed because the "cold equations" of spaceflight determine that the ship cannot reach its destination and deliver its supplies because it lacks the fuel and oxygen reserves to transport her as well as the pilot. The argument is one of technological determinism: the "laws" of space dictate a murderous ruthlessness of autocratic governance by the pilot. Doctorow's (2014) analysis, however, shreds the argument, claiming that, as "design fiction" for a sociotechnical system of spacecraft and governance rules, it enables "moral hazard." In short, the fictional system was designed to give rise to the political outcome, and could easily have been designed otherwise. Rather, the system was built to enable a particular set of values in action, leading to an immoral, tyrannical outcome. An alternative system could have been constructed with autopilots, emergency reserves—not to mention capabilities of synthesizing medicine within the colony itself. Thus, whether the sociotechnical systems at issue are in space or online, an analysis of their political implications must question whether a particular configuration is unavoidable, or whether alternative systems embodying a different set of values might be created.

The second point is closely related: infrastructural systems are designed in accordance with existing cultural values. They then tend to reproduce those values in use. These systems are used by people who are similarly the product of existing values: even given a desire to innovate or to reject an established order, they still bring it with them. Scholars from de Toqueville (1835, 1840 [2003]) to Turner (1893) to Foucault (1967) have observed that social change in "new" spaces is an emergent property of interaction of persons laden with cultural baggage and a new physical and sociotechnical environment: there are no utopian breaks, no

"Year Zero"s, but a dynamic synthesis of old and new. One implication of these two points should be obvious, and Foucault (1967) makes it relatively explicit: the extent to which settlement infrastructure embodies the values of the metropolis, the settlers continue to adhere to those values, and the environment permits, little will change, while the converse is also true.

One clear implication here is that governance systems may take a variety of forms, and that any particular problem of governance may be solved by culture, law, or engineering. "The Cold Equations" presents an engineering solution to a social problem of stowaways. Doctorow (2014) shows how that social problem might have several alternative engineering solutions. The story shows a legal solution—a "Do Not Enter" sign—as having failed. One could imagine a cultural solution, a strong social prohibition against stowing away, or alternatively, a software-engineering solution—in which an excess-weight reading would prevent takeoff. Engineers tend strongly to prefer engineering solutions to social problems: many of the "design fictions" (Doctorow 2014) described above take such a perspective, and a preference for software solutions to social problems is one of the defining characteristics of the "Web 2.0" (O'Reilly 2005) era. The case of virtual worlds presented below epitomizes the clash between cultural and engineering solutions to core issues of governance, reflecting a divergence of values between the engineers designing the infrastructure and the users inhabiting the social space it created.

Thus, infrastructure reflects a choice among a range of different values, and is rarely fixed into a single possible configuration by laws of physics. That configuration tends to perpetuate those values. People are capable of choosing among the values of the culture that raised them, but cannot escape them by wiping their social, ideological, and moral slates clean. New values arise from a complex and highly contingent set of interactions among technological infrastructure, cultural infrastructure, physical environment, and values or vision. Taken together, these statements suggest that any assemblage of environment, infrastructure, and culture can be "read" for its prospect of promoting liberty, tyranny, or social cohesion, and that modifications to any element of the assemblage may alter that prospect in a nonlinear fashion. With that in mind, we can now analyze the case for virtual worlds both as a libertarian space and as an analog to extraterrestrial polities.

8.5 Testing the Space-Libertarian Hypothesis: Virtual Worlds as Case Study

The spread of internet access led some to claim that "virtual communities," a term coined by Howard Rheingold in his study of the pioneering group The WELL in 1993, (Rheingold 2000) would pioneer newer, more democratic political forms as Barlow envisioned. This vision was shared by many, both developers and users, in the virtual world of SL, which launched in 2003. (see, e.g., Ondrejka 2007)

However, the very few experiments in democracy undertaken in SL proved vastly less popular as communities and exemplars of governance than its founders expected.

As noted above, SL was closely linked to the techno-libertarian views espoused by the Electronic Frontier Foundation, among other organizations. It was envisioned as a digital analog to the Burning Man festival: a place for creative engineers and designers to show off their artistry in an environment with a minimum of enforced rules and a spirit of anarchic play (Au 2008; Malaby 2009). Au, hired by SL's parent firm LL as an "embedded journalist," and later principal of virtual worlds' main news blog, provocatively entitled *New World Notes*, explicitly cites Nozick (1974) in his depiction of the early SL social environment: one free of social rules, in which neighbors either got along or went to war, in which universities' virtual campuses, historical military re-enactors, brothels and art galleries might find themselves adjacent, with patrons freely crossing each other's properties.

It was into this environment that I entered for a 5-year doctoral research project: ethnographic fieldwork in a range of communities experimenting with new forms of governance. With a primary mission of exploring the adaptation of internet technologies generally, and persistent, avatarized virtual worlds specifically to matters of governance, I also saw the project as able to shed light on socio-political behavior in a transhumanist environment, in which material scarcity and bodily needs would be obviated by technology, given a profound interest in these spaces by transhumanists building on Barlow's vision of a republic of the Mind. Somewhat less speculatively, if less closely modeled, I saw the potential to apply conclusions from the study to concepts of extraterrestrial governance. Virtual worlds had the potential to model certain features of an O'Neill/Nozick vision of space settlement, in which large numbers of people (in the mid-to late 2000s, SL had between 40 and 100,000 "residents," or users, on at any one time, day or night) sought, in Barlow's terms (picked up and applied to virtual worlds by the economist Edward Castronova, who saw these spaces as out-competing the physical world by offering greater personal value (Castronova 2007)) to leave the status quo of terrestrial life for a "new frontier" of personal reinvention. Unlike national and local governmental experiments with "e-democracy," virtual worlds offered relatively free flows of value, information, and people; potential new tools for decisionmaking and resource allocation; a lack of legacy systems, and, again, an explicit techno-libertarian design. This paper summarizes that work: a case study (McKnight 2012) and my dissertation (McKnight 2013) provide a more complete picture.

The virtual world of SL is a persistent simulated three-dimensional environment in which people interact via avatars, or graphical representations of themselves (which need not reflect their physical form in any way). Unlike typical massively multiplayer online games (MMOs) such as World of Warcraft, it is not built around game mechanisms of quests, scores, competition or objectives. Rather, it re-creates a physical space much like the real world, in which people engage in such activities as they find valuable. Personal behavior within SL is governed by a set of contracts between the user and LL, to which users enter into by clicking a button indicating assent, which must be done in order to enter into the virtual world (Lastowka 2010).

Among these is the "Community Standards" document prohibiting a "Big Six" of behaviors: expressions of intolerance, harassment, assault, disclosure of real-life information, violation of the adult content zoning, and disturbing the peace. Users are told that they "should report violations" using an Abuse Reporter tool built into the viewer software (Linden Lab 2011). With approximately 1.5 million regular monthly users and an average of 50,000 people "in world" at any given time, (Nino 2011) the general perception that LL does not swiftly and firmly investigate abuse reports and sanction violators would seem to be justified.

As LL disclaims liability for "Content, conduct or services of users or third parties," (Linden Lab 2011) a user has no claim against LL for not sanctioning any particular alleged violators of the Terms of Service or Community Standards: there is no basis for "third-party beneficiary" claims relating to the contract between LL and the alleged violator, (Fairfield 2008) and no basis in the contract between LL and the person alleging violations, as LL has no contractual duty to act on abuse reports, and specifically disclaims liability for actions of other users. This contractual situation gives rise to a political vacuum: not only is the abuse reporting system practically ineffective, particularly in real time, there is also no mechanism for resolving disputes between users that do *not* involve allegations of violations of the Terms of Service. Users have no contractual relationship with each other, and there is no body of tort law, no judicial system, no universally agreed-upon means of dispute resolution, and most importantly, no mechanism for enforcing judgments, within SL.

SL users have expanded upon their land-ownership powers as a means of filling this vacuum. While "mainland" land, owned directly by LL, bears no restrictions on its use other than those relating to mature content guidelines, the bulk of land in SL is composed of private islands, typically subject to a "covenant," a document setting forth terms of rental, the sorts of limitations on uses and appearance common in real-life homeowner agreements in planned communities, including architectural style, limitations on the location of unfinished projects, additional restrictions on speech, and other terms. Additionally, much of the mainland has been purchased by "land barons," or large leasing companies, which apply covenants and then sublease the land for residential or business uses. Land barons, or their management staff to whom land powers have been delegated, are typically much quicker to respond to complaints and disputes on their properties. With the ability to evict and ban sublessors, land barons have actual enforcement authority over their tenants.

The land baron/covenant model arose with the establishment of private islands as an SL product in 2006. From SL's launch in 2003, however, there was no clear solution to the lack of a dispute resolution system within the structure of SL's software and LL's contracts with its users. In 2004, a thread on the SL official forum hosted a discussion of the prospect of users creating institutions to fill the vacuum left by LL. In response to a call for proposals by a Linden employee, a group of the forum members submitted a proposal to build a community to be managed pursuant to an electoral system, with a constitution and provisional government. At the time, participants hoped that their model of a constitutional, elected government would prove popular enough to spread to 5 % of the SL grid. Instead, the grandly-entitled

Confederation of Democratic Simulators stabilized at five regions, or "simulators," each representing 65,536 square meters of virtual land, out of a total of 31,426 regions as of April 2011, (Shepherd 2011) and 70 voting members out of SL's regular user base of 1.5 million.

My 5-year study of SL found only four other experiments with democratic self-governance in SL: one community populated by transhumanists and space exploration advocates (which held one vote, to abolish the democratic experiment and transition to a managerial model: only the managers voted against abolition), one Italian-language artists' cooperative, a small community run by American educators, and a group created to enable dialog between secular Westerners and Muslims from around the world, which merged into the CDS, then a year later separated from it and subsequently collapsed. In total about 600 people of the millions who established avatars in SL participated in one or more of the governance experiments.

In MMOs, the situation was roughly similar: the almost universal form of organization was the pyramidal-structured guild (or equivalent such as "fellowship" in Lord of the Rings Online or "fleet" in Star Trek Online: the names reflected the world's narrative, but organizational charts were substantially identical). Typically the guild leader was all-powerful, the software of the MMO prohibiting the delegation of powers even when users might have sought a less hierarchical structure. While MMOs were often viewed from the outside as a chance to act out fantasies of power, in practice they provided an escape from responsibility: by far the most difficult positions for guilds to fill were those at the top of the hierarchy, as leadership was generally seen as an unremunerative burden and the antithesis of fun. Software solutions to social problems were commonplace, one of the most creative and popular being a user-developed technology called "Dragon Kill Points" as a means for redistribution akin to taxation and social welfare policies (Castronova and Fairfield 2007). While MMOs saw innovations in the development of software solutions to socio-political problems, they never faced calls for liberty in a Barlow sense. Hierarchy prevailed, and competed in a liquid market for low-level followers, who found themselves in an almost ironic version of Nozick's libertarian utopia: myriad governments in which barriers to entry and exit were low, such that an active free market in government could take place, but one in which only one type of government, the feudal/corporate hierarchy, was on offer.

Generally, virtual worlds presented a political atmosphere much like those of Heinlein's Lunar Authority or some sorts of colonial regimes: an overlord with technically near-unlimited power, which however chose not to intervene in internal disputes among residents. This political vacuum at the local scale has been seen by novelists and political theorists alike as an ideal environment for innovation and experimentation, based on the assumption that people, especially those drawn to unsettled new environments, would seek to maximize their liberty. The case of virtual worlds, however, demonstrated the opposite: what people almost universally chose was familiarity and delegation of responsibility, via hierarchical and managerial systems.

Putnam (2001) extensively documents the decline of associational participation among Americans after 1960; Warren (2000) observes the strong consensus among political scientists that, *pace* Tocqueville, the strength of a democracy lies in the "robustness of its associational life." This line of thinking holds that, through American history from Colonial days until the mid-1960s, American democracy was a skill learned in practice—not, in the first instance, in local government, but in the mutual governance of voluntary associations, from religious to political to recreational. It was in these groups that Americans learned parliamentary procedure, negotiation, compromise, the arts of campaigning and governing. Some would choose to enter into the formal institutions of government; most would have experiential knowledge by which to judge the performance of those who did. At a more fundamental level, participatory self-governance was reinforced as a norm, both expectation and obligation, which then extended into more state-focused participation such as voting and jury duty.

Putnam documents the extensive transformation in American participation in the period from roughly 1965–1985; he ends by speculating that only the new realm of online communities might counter the trend of decline. We saw above that early internet activists shared Putnam's view: Rheingold (2000) and Turner (2008) trace a desire to reinvent participatory civic culture from the hippie commune movement directly—via a handful of common activists—to early internet culture. By way of the Electronic Frontier Foundation and Burning Man, SL is in direct lineal descent of this Bay Area-based movement to reimagine the egalitarian construction of new forms of community to serve the old ends of building democracy through practice. Yet SL only produced a small handful of participatory communities, and only one with an elective, albeit entrenched, leadership cadre. The experience of those communities answers why: while there were some democratic aspirations, there were profoundly few practicing democrats.

Even among a population seeking community—it was after all one of the few things SL affirmatively offered to retail, as opposed to corporate and educational, customers—both will and skill were so rare as to have been found in a bare handful of persons in SL's middle years. There was no shortage of highly skilled designers of retail goods: SL's fashion creators generated the vast bulk of $1million a day in peer-to-peer transactions among SL users throughout the period. Likewise, millionaire land barons made the cover of Business Week magazine (Hof 2006). The only thing in short supply was people willing to provide the infrastructure of fun.

Thus, even as SL's demographic skewed heavily towards those whose California Ideology most predisposed them to seek virtual community (Au 2008) what proved missing was not only the how-to of self-governance experience, but a concept of fun which included the obligations of management of the social. The same mechanism underlies the popularity of planned communities in the once-frontier West: they provide the reassuring framing of pre-mass community while outsourcing the responsibilities to corporate management. This is also symptomatic of a sort of democracy in which the symbols of an age of popular self-governance are bought and sold, but actual practices are outsourced willingly by a populace trained to think of governance as distasteful and worth paying to be rid of. As the population drawn to virtual worlds was substantially similar demographically to those drawn to managed communities in the physical world, the parallels should have been expected. Likewise, a similar demographic—English speaking, highly technologically literate, skilled workers, with a desire to be early adopters of new technologies—will be seen in any future era of space settlement. Extrapolation of the actual governance practices seen in virtual worlds to the case of space settlement thus seems justified.

8.6 Modeling Space Governance Systems: Where Can We Go from Here?

The case of virtual worlds provides one salient data point for considering the forms of governance which may emerge or be supported in future extra-terrestrial settlement. Substantial work is being done in simulating the social environment of early small-scale deep-space and Mars missions (e.g., the NASA Haughton Mars Project, [30] the Mars Desert Research Station, [31] the Hawai'i Space Exploration Analog and Simulation, [32] and Mars 500 [33]), which builds on decades of extreme environment social research. However, with Mars settlement actively proposed and debated, research into socio-cultural issues of permanent extra-terrestrial settlement is no longer premature. As the remote Burning Man festival was viewed as a model by the designers of SL, (Au 2008) we may look to extant and custom-built physical communities as models for extra-terrestrial settlement: Arizona's Arcosanti and Biosphere 2 have been useful experiments in that regard. Beyond the sort of ethnographic work in virtual worlds which I have performed, custom platforms may be created for social science experimentation, as economist Edward Castronova has long advocated (Castronova 2005).

As Malaby (2009) documents and my own work supported, virtual worlds saw a socio-cultural disconnect between designers and users, which had significant negative ramifications on governance and culture. Designers—predominantly software engineers—valued technical skills, a do-it-yourself ethic, and expected the users of their virtual-world platforms to be similarly motivated. Users, however, tended to value aesthetics, social interaction, consumerism, and conspicuous consumption/status display. Much of the story of designer/user interaction within virtual worlds has grown out of the resultant clash of values and practices. The potential for a similar conflict to arise between aerospace engineers and off-planet emigrants is high. Modeling, simulation, and ongoing social science research may aid in shaping the design process to prevent or mitigate clashes over values, technologies and processes, and may make the difference between the success or failure of a future settlement.

This story of the mis-fit between the techno-libertarian ideals of entrepreneurs and engineers on the one hand and early adopters of social technologies on the other is a cautionary tale in at least one additional respect: besides suggesting the need for building lessons learned into the design process, it suggests that there may be a fundamental disconnect between the types of people likely to build and inhabit space settlements and the values and practices most likely to produce sustainable societies in those habitats. In that respect, Marsden's examination of Indigenous practices for long-term sustainability (Marsden 2013) may prove crucial: drawing on the experiences of those cultures which have survived longest in extreme environments may be the key to reshaping space settler's cultures so as not to propagate a dynamic shown to be problematic in analogous environments (a point also made by Cockell 2013, p. 70, n. 7).

It may be that rather than a shortage of space libertarians to inhabit the envisioned spaces of advocates and entrepreneurs, we may instead choose to address a shortage of space communitarians, and turn to a set of best practices honed over millennia by Indigenous communities around the world as inspiration for our future settlements beyond the Earth. Such an approach breaks with a false dichotomy between liberty and tyranny common in a line of political reasoning derived from the European Enlightenment by using both "design fictions" (Doctorow 2014) and field experiments to explore systematically the interplay of ideology and engineering in creating infrastructures which make particular social outcomes more or less likely, cultural predispositions and utopian ideals, and environmental specifics in giving rise to new physical and social technologies capable of enabling humans to flourishin extraterrestrial environments.

References

Au, W. J. (2008). *The making of second life: Notes from the new world.* New York: HarperBusiness.

Barlow, J. P. (1996). *A declaration of the independence of cyberspace.* https://w2.eff.org/Censorship/Internet_censorship_bills/barlow_0296.declaration.

Barbrook, R., & Cameron, A. (2007). The Californian ideology. *The HRC Review.* http://www.imaginaryfutures.net/2007/04/17/the-californian-ideology-2/.

Billings, L. (2007). Overview: Ideology, advocacy, and spaceflight–evolution of a cultural narrative. In S. J. Dick (Ed.), *Societal impact of spaceflight* (pp. 483–499). Government Printing Office.

Bookchin, M. (1991). *Libertarian municipalism: An overview. Green perspectives,* No. 24. Burlington, VT.

Castronova, E. (2005). *Synthetic worlds: The business and culture of online games.* Chicago: University of Chicago Press.

Castronova, E. (2007). *Exodus to the virtual world: How online fun is changing reality.* New York: Palgrave McMillan.

Castronova, E., & Fairfield, J. (2007). *Dragon kill points: A summary whitepaper.* SSRN 958945.

Cherny, L. (1999). Conversation and community: Chat in a virtual world. CITY: Center for the Study of Language and Information.

Cockell, C. (2013). *Extra-terrestrial liberty: An enquiry into the nature and causes of tyrannical government beyond the Earth.* Edinburgh: Shoving Leopard.

de Toqueville, A. (1835, 1840 [2003]). *Democracy in America.* New York: Penguin.

Dibbell, J. (1999) My tiny life: Crime and passion in a virtual world. New York: Holt Paperbacks.
Doctorow, C. (2014). Cold equations and moral hazard. *Locus Online*. http://www.locusmag.com/Perspectives/2014/03/cory-doctorow-cold-equations-and-moral-hazard/.
Dubner, S.J. (2007). "Philip Rosedale answers all your Second Life questions." http://www.freakonomics.com/2007/12/13/philip-rosedale-answers-your-second-life-questions/
Fairfield, J. A. T. (2008). Anti-social contracts: The contractual governance of virtual worlds. 53 McGill L.J. 427.
Foucault, M. (1967). *Of other spaces*. http://foucault.info/documents/heteroTopia/foucault.heteroTopia.en.html.
Foust, J. (2010). SpaceShipOne, government one? *The Space Review*.http://www.thespacereview.com/article/1649/1
Gartner (2007). "Gartner report says 80 percent of active internet users will have an active 'Second Life' in the virtual world by the end of 2011." April 24, 2007. http://www.gartner.com/it/page.jsp?id=503861
Godwin, T. (1954 [2003]). *The cold equations and other stories*. New York: Baen Books.
Gibson, W. (1982). "Burning chrome." Omni Magazine, July 1982–1984. Neuromancer. New York: Ace Science Fiction.
Heinlein, R. A. (1966). *The moon is a harsh mistress*. New York: G. P. Putnam's Sons.
Heinlein, R. A. (1985). *The cat who walks through walls*. New York: Putnam Publishing Group.
Kapor, M. (2008). "Closing Keynote." http://wiki.secondlife.com/wiki/SL5B/Transcripts.
Lastowka, G. (2010). *Virtual justice: the new laws of online worlds*. New Haven: Yale University Press.
Limerick, P. N. (2000). *Something in the soil: Legacies and reckonings in the New West*. New York: W.W. Norton & Co.
Linden Lab. Community standards. Retrieved April 8, 2011, from http://secondlife.com/corporate/cs.php.
Locke, J. (1689). *Two treatises on government*. http://socserv2.socsci.mcmaster.ca/econ/ugcm/3ll3/locke/government.pdf.
Ludlow, P. (Ed.). (1996). *High noon on the electronic frontier: Conceptual issues in cyberspace*. Cambridge: The MIT Press.
Malaby, T. M. (2009). *Making virtual worlds: Linden lab and second life*. Ithaca: Cornell University Press.
Marsden, D. (2013). Indigenous principles for a starship citizen handbook. *Proceedings of the 2013 100 Year Starship Public Symposium*.
McCurdy, H. (1997). *Space and the American imagination*. Washington: Smithsonian Institution.
McKnight, J. C. (2012). A failure of *convivencia*: Democracy and discourse conflicts in a virtual government. *Bulletin of Science, Technology and Society, 32*(5), 359–372.
McKnight, J. C. (2013). *The resilience engine: Generating personhood, place and power in virtual worlds, 2008–2010* (Doctoral dissertation, Arizona State University).
Morningstar, C. and Farmer, F.R. (1991). "The lessons of Lucasfilm's Habitat." In Cyberspace: First Steps, Michael Benedikt (ed.), 1991, Cambridge: The MIT Press.
National Commission on Space. (1986). *Pioneering the space frontier: Report of the National Commission on Space*. New York: Bantam Books.
Nino, T. (2011). Second life statistical charts. Retrieved April 12, 2011, from http://dwellonit.taterunino.net/sl-statistical-charts/.
Nozick, R. (1974). *Anarchy, state and utopia*. New York: Basic Books.
O'Neill, G. K. (1976). *The high frontier*. New York: William Morrow and Company.
O'Reilly, T. (2005) What is Web 2.0: Design patterns and business models for the next generation of software. Retrieved September 30, 2005, from http://oreilly.com/web2/archive/what-is-web-20.html.
Ondrejka, C. (2007, Summer) Collapsing geography: Second life, innovation and the future of national power. *Innovations*.
Polyani, M. (1967). The republic of science: Its political and economic theory. *Minerva, 1*(54–74), 1962.

Rheingold, H. (2000). *The virtual community: Homesteading on the electronic frontier.* Cambridge: The MIT Press.

Robinson, K. S. (1992). *Red Mars.* New York: Bantam Spectra.

Robinson, K. S. (1993). *Green Mars.* New York: Bantam Spectra.

Robinson, K. S. (1996). *Blue Mars.* New York: Bantam Spectra.

Robinson, K. S. (1997). *Antarctica.* New York: Bantam Spectra.

Shepherd, T. (2011). Second Life grid survey. Retrieved April 14, 2011, from http://www.gridsurvey.com/.

Sydel, L. (2010). "Sci-fi inspires engineers to build our future." National Public Radio. http://www.npr.org/templates/story/story.php?storyId=129333703

Turkle, S. (1997) Life on the screen: Identity in the age of the internet. New York: Simon and Schuster.

Turner, F. (2008). *From counterculture to cyberculture: Stewart Brand, the whole Earth network, and the rise of digital utopianism.* Chicago: University of Chicago Press.

Turner, F. J. (1893). The significance of the frontier in American history. In J. M. Faragher (Ed.), (1994) *Rereading Frederick Jackson Turner.* New Haven: Yale University Press.

Williamson, J., & Breuer, M. J. (1981 [1931]). *Birth of a new republic.* Enterprises: P.D.A.

Winner, L. (1980). Do artifacts have politics? *Daedalus, 9*(1).

Zubrin, R. (1994). The significance of the Martian frontier. Reprinted in Zubrin, R. (1996). *The case for Mars: The plan to settle the Red Planet and why we must.* New York: Simon and Schuster.

Chapter 9
Citizens of Mars Ltd.

Erik Persson

Abstract When the time comes to decide how to govern an extraterrestrial settlement there will be many alternatives to chose from. We will have the opportunity to try new and so far untested theories, but there are also some old forms of government that might be tempting to try again. We might for instance let the company whose activities on the world are the reason for the establishment govern the settlement. This has been tried before on our own planet both because it was seen as convenient and as an incentive for colonisation. In this chapter I will ask what this solution would mean for the civil liberty of the settlers. To answer the question I will look at some historical analogues and have a philosophical discussion. The conclusion is that a settlement governed by a body whose sole reason for existence is to make money for the owners, that is led by a board that answers only to the owners and not to the people, that functions as both government and sole employer, and that has the unlimited power over the life support systems necessary for the survival of the settlers will not be a good basis for civil liberties.

Keywords Extraterrestrial liberty · Economic liberty · Civil liberties · Corporation government · Space commercialisation

9.1 Introduction

One of the things that seem the most fascinating about establishing a new self-governing society in space is that we would get a chance to do it from scratch. We could take the chance to try out ideas that for us who live on the earth today would be just utopias.

On our own planet we already have societies. All societies already have a history, a culture and some kind of political system. All societies have people with

E. Persson (✉)
Lund University, Lund, Sweden
e-mail: Erik.persson@fil.lu.se

C.S. Cockell (ed.), *Human Governance Beyond Earth*,
Space and Society, DOI 10.1007/978-3-319-18063-2_9

power who get their power from the existing system and who are prepared to use that power to preserve the system. This makes it very difficult to try something radically new. Revolutions all too often turn over the power to people who are good at leading revolutions but are not necessarily as good at leading a country (or particularly well versed in political philosophy). Revolutions are also notoriously bloody affairs. Having access to a completely "new" world where one could construct the ultimate political system without having to overthrow any existing power structures would provide a unique opportunity to try out new ideas in a more peaceful way.

We have to remember, however, that even a new civilisation on another, previously uninhabited world, will have a history. That is, it will have our history and the history, still to come, of how that world came to be colonized by us. This history will of course strongly influence what kind of society it will become. It would not be too far-fetched to assume that a new world colonized by a country on the earth would, at least initially, be governed by that country. The same would probably be the case if the new world is colonized by a group of countries. If it for instance will be colonized by the European Space Agency, it is a fair guess that it will be governed by the EU. If the new world will be colonized by a truly global effort it might be governed by the UN.

How about if it is colonized by a corporation? Could it then also be ruled by a corporation?

It has been argued that an increased economic liberty in space will be necessary to create the incentive needed for creating extraterrestrial settlements on other worlds (see e.g. Cockell 2015; Reinstein 1999; Simberg 2012). Even in a situation where corporations that operate on another world are granted a very high degree of freedom, the society made up by the workers and their families needs to be governed in some way. It will need laws, law makers and courts. How can this be accomplished if we do not want the settlement to become a colony of an earth nation? The ultimate solution would be that the settlers govern themselves. Another solution that might be tempting to some could be to give the corporation that operates on the word in question the right to rule over the settlement. Handing out the right to rule as an incentive for exploration and exploitation has been tried before. In the 15th century just after Christopher Columbus' discovery of the new world, the British king Henry VII in an attempt to take up the race with Spain and Portugal offered explorers, not just a monopoly on the trade in any land he found, but also the right to rule over this land (Newman 1985).

Letting corporations rule of over newly colonized worlds on which they operate might have certain advantages. It would in particular be a radical way of preserving the settlement's independence from earth governments, and the freedom of the corporation from regulations set up by earth governments, but how would it affect the personal freedom, also known as the civil liberty, of the inhabitants?

Extraterrestrial liberty can be interpreted as meaning simply that extraterrestrial settlements are self-governing in some way. It can, however, also mean something more. It can mean that future inhabitants of extraterrestrial settlements enjoy a substantial degree of personal freedom in relation to their government (civil liberty), or it can mean that corporations operating off the earth enjoy a large degree of freedom from interference by earth governments (economic liberty). Can and should increased economic liberty in space imply that corporations get the right to establish their own states? Would a corporation acting as a government be the pinnacle of economic liberty or a nightmare for civil liberty, or both?

In this chapter I will discuss the implications for the civil liberties of individual extraterrestrial settlers in the hypothetical, but far from implausible, scenario where a settlement is governed by a corporation.

9.2 Commercialisation of Space

Since the end of the Apollo era, the motivation for governments to engage in space exploration has cooled down considerably. It has therefore become common to claim that an increased commercial interest in space is needed in order to speed up our entering of the space age (Cockell 2015; Collins 2008; Meyer 2010; Reinstein 1999; Simberg 2012). Proponents of human colonisation of space therefore often propose to give more freedom to the market powers in space (See e.g. Collins 2008; Reinstein 1999; Simberg 2012). In other words, they want more economic liberty in space. Some are also quite optimistic about the possibilities to actually make money from space primarily from mining and tourism (Beery 2012; Collins 2008; Meyer 2010; Reinstein 1999; see also Deep Space Industries 2014; Moon Express 2014; Planetary Resources 2014; Schackleton Energy 2014; Spaceport America 2014; Spaceport Sweden 2014 for examples).

The commercial involvement in space exploration has, in fact, been prevalent since the beginning and it is steadily increasing (Beavin 2008; Beery 2012; Collins and Ashford 1988; DFI International 2002; Ehrenfreund et al. 2013; Giacalone 2008; Hertzfeld 2002; Hubbard et al. 2013; Lewicki et al. 2013; Meyer 2010; Murthi et al. 2007; NASA HQ Library 2014; NASA Policy Directive 7410.3 J; NASA Policy Directive 8610.12G; Peeters 2002, 2003; Persson 2014; Reinstein 1999; Velocci 2012). There is also a pronounced wish from the US government, that the commercial involvement increases even further (Bush 2002; Beavin 2008; Giacalone 2008; U.S. Department of Commerce 2014; U.S. Federal Aviation Administration 2014).

This trend seems to strengthen the suspicion that the initiative and money behind the first extraterrestrial settlements might well come from commercial interests and not from earth governments. This in turn does not have to imply that the first human settlements in space will also be governed by the companies behind the settlements, but it is not implausible that they will. During the colonial era, colonies on our own planet were often governed by corporations (Stern 2011). Placing the governing power over an extraterrestrial society in the hands of a company that operates on the world in question might in fact be very convenient. A company operating on another world already has command structures, administrative structures, personnel, infrastructure, etc. in place. To set up and pay for an organisation of civil servants to administer the colony on behalf of an earth government will be expensive and not necessarily politically attractive for the earth government. Letting companies that operate in space form their own states would also be a way of guaranteeing political independence from the earth and to increase the freedom for the company to operate without being hampered by nationalistic motives, and without having to navigate its way through an intricate web of national and international law.

Maybe increased economic liberty in space, including the right for corporations to form their own societies in space, is the right way to go if we want something to happen fast. I am not going to discuss that question here though. Instead, I will just take it as a plausible scenario that at least some of the first space settlements might be governed by corporations, and discuss how this will affect the civil liberty of the space settlers if it happens.

9.3 Economic Liberty and Civil Liberty

In order to answer our question, we need to say a few words about two kinds of liberty: economic liberty and civil liberty.

Economic liberty usually refers to a relative absence from state regulations of the economic markets, while *Civil liberty* refers to freedom of the individual from oppression by the state. *Civil liberties* are liberties of the individual that are derived from or needed to uphold civil liberty. Some examples of civil liberties found in the literature are: the right to life and personal safety (Gearty 2007; Skaaning 2009; UN 1966), freedom of thought and speech/expression (Dobkin 1976; Gearty 2007; Mill 1983; Shi and Lou 2010; Skaaning 2009; Sturm and Wright 1976; UN 1966; Wasby 1976), the right to peaceful protest (Des Rosiers 2012), freedom of association/assembly (Dobkin 1976; Gearty 2007; Skaaning 2009; UN 1966; Wasby 1976), freedom of religion (Dobkin 1976; Mill 1983; Shi and Lou 2010; Skaaning 2009; Sturm and Wright 1976; UN 1966; Wasby 1976), the right to a fair trial/due process (Dobkin 1976; Gearty 2007; Mill 1983; Skaaning 2009; Sturm and Wright 1976; UN 1966), and the right to be part of the political process and in particular in the appointment of government (e.g. in the form of the right to vote) (Gearty 2007; Locke 1988; Skaaning 2009; Sturm and Wright 1976; UN 1966).[1]

It will not be necessary to delve into the definitions in much more detail than this. The exact definitions of the two terms can be debated at length. That is, however, not what I will do here since it would bring us too far from the main question. There are also other categories of liberties than the two mentioned above. It will for the same reason not be necessary to bring them up here.

A few explanatory sentences about the relation between economic liberty and civil liberty in the case we are discussing are warranted though.

One might be tempted to claim that economic liberty is an element of civil liberty or at least that the two are so closely tied together that economic liberty

---

[1]For a list of civil liberties indexes see e.g. Skaaning (2009).

cannot be at odds with civil liberty. Both liberties are after all about limitations of the power of the state.

Given the way economic liberty and civil liberty are described above it still makes sense to keep them apart, though. Given the case we are discussing where economic liberty includes the right of corporations to form their own states, it also seems clear that in this particular case, if not in general, economic and civil liberty can be in opposition.

Economic liberty is about freedom of markets while civil liberty is about freedom for individuals (cf. Skaaning 2009). It is true that individuals are agents on markets but so are corporations and even states (even when the state is not a corporation). Civil liberties on the other hand can per definition only be granted to individuals. A liberty that includes corporations and states among its objects can therefore not fit into the concept of civil liberty.

This is particularly important for our discussion since what we discuss here is a very special case where a corporation *is* the state. This means that in this special case, the issue of civil liberty of the individuals in this society will in fact be a matter of the extension of the power of the corporation over the individual settlers. This creates a very special situation that demands special analysis and that motivates that we keep the two types of liberty separate.

It might seem odd or even contradictive that economic liberty includes states as its objects considering that it implies a limitation of the power of states. We are talking about two different things, however. The limitation of the power of states concerns their power to regulate markets. This does not limit their liberty to be actors on the market as long as they play by the same rules as every other actor.

That economic liberty can be extended to allow corporations to act as states might also seem confounding considering what we said above. What makes it possible is again the role of states as regulators. Nothing stops recognised states on the earth to agree to hand over the right to form a state on another world to the market powers. When that is done, however, and a corporation has taken on the role as government, the concept of economic liberty in that particular state cannot be applicable to the ruling corporation since that is the state. One can also suspect that such an arrangement will be detrimental to the economic liberty of other corporations on this world. This is, however a different question. Let us get back to the issue of how it affects the civil liberties of the individuals, and let us start by a look in the rear view mirror.

9.4 Civil Liberties in Colonies Ruled by Corporations in Earth History

One way of approaching the question is to look at the colonial history on our own planet.

People usually think of colonial powers as nations with colonies, like the British Empire. It was not uncommon, however, that corporations were in charge, not just of the trade but of all aspects of life in a colony, including the right to decide and enforce local laws and to judge in local matters (Nelson 2008; Newman 1989; Payne 1989; Stern 2011).

The East India Companies are the most well known examples of this system. Several countries in Europe had them, but the British one was by far the most powerful. "The Governor and company of merchants of London trading into the East Indies" as its original name read, was for centuries the facto rulers over all British territory in Asia and both its native population and any European settlers (Stern 2011).

The East India Companies were not unique in this respect. The Hudson's Bay Company (HBC)—or as its full name read: "The Governor and Company of Adventurers of England, Tradeing into Hudson's Bay" (from the company charter signed 1670 by Charles II, reprinted in Newman 1985), ruled over the entire drainage area of the Hudson's Bay basin (Carlos and Lewis 2010). The British South Africa Company ruled over Northern Rhodesia for over 30 years during the 19th and 20th centuries (Slinn 1971), and the Virginia Company of London founded and ruled over Virginia in what is now USA until economic problems forced them to accept other investors and as a result thereof to accept English law (Nelson 2008). There were many more examples but let us limit ourselves to these four in order to keep our focus.

What was the state of civil liberties in these areas?

For the native population, the circumstances seem to have varied. In many areas, the local resources were very heavily exploited, which can be seen as a violation of the locals' property rights but also of their right to life since the exploitation in some cases threatened their subsistence. The beaver in the Hudson Bay area was hunted almost to extinction (Carlos and Lewis 2010; Payne 1989). In India, the monopoly on the rice trade bankrupted local factories and caused famine among the population (Tunick 2006).

Apart from the fur trade, the interventions by the HBC in the lives of the locals, seem to have been limited to occasional peace keeping missions (see e.g. Carlos and Lewis 2010; Payne 1989). Apparently they did not make any serious missionary activity (Payne 1989). It happened at some occasions that representatives of the company administered punishments to the natives but it was considered imprudent to do so and most of the time, the company officials abstained from that kind of intervention (Payne 1989).

In Northern Rhodesia there were apparently no big uprisings by the natives against the British South Africa Company (Slinn 1971). A possible explanation for this can be found in the use of indigenous leaders as administrators (Slinn 1971). The use of native administrators might have taken some of the heat off from the company and might also have meant that everyday life for the majority of the native population was not that different from what it used to be like before the colonisation.

The British and French East India Companies also relied on cooperation with indigenous leaders, apparently mainly because it was considered too expensive to have paid European officials to do the work (Gren 1969). The company also relied heavily on alliances with local leaders (Stern 2011). The local leaders were called "petty kings" by the company and their titles were apparently mostly ceremonial and used to pretend that the locals had influence over how the territory was governed (Stern 2011). These measures did not prevent several uprisings to occur, however (Gren 1969; Stern 2011).

Using locals as administrators will for obvious reasons not be an option in our scenario, however, since as long as we talk about our own solar system, we can be quite sure that even if we find life we will not find any existing civilisations outside our own earth.

A very clear infringement of civil liberty practiced by the British East India Company was the use of native slaves. Even though the company did not allow their officials to use slaves privately, slaves were used by the company itself, and they were also presented as gifts to local leaders (Stern 2011). The punishment of slaves was described as very hard (Stern 2011) though it is not clear whether it was more severe than the punishment of slaves under other forms of government.

A particularly aggravating fact was that company courts sometimes sentenced people to become slaves for the company, which seems like a rather blatant misuse of power.

It should be pointed out that the company did not impose the practice of slavery on the native populations. They just continued an already established tradition and therefore did not impose any new curtailment of civil liberty. This cannot excuse the use of slavery but it shows that in that particular respect, company rule might not have been worse than alternative forms of government in that area at that time.

Company law incorporated both English law and local laws and traditions (Stern 2011). This was one of the methods practiced for keeping good relations to the native populations, but it also enabled the company to cherry pick the traditions that suited them best, such as the local tradition of sentencing people to slavery (Stern 2011) and the Indian tax system that provided the company with much higher tax revenues than the British system would have done (Marshall 1997). In one respect the company rule was in fact an improvement in that they banned the export of children as slaves (Stern 2011).

When it comes to the question of due process and fair trials, it can be noted that it was possible for natives to win in court against company employees (Stern 2011). I have not found any case where a native have won a case against the company as such, however.

The matter of religious freedom was apparently complicated under the rule of the British East India Company. On one hand, they were compelled to promote or at least protect Protestantism and to limit the influence of other religions, especially since religion played a vital role in the justification of their ruling. They also received considerable pressure from England to increase their efforts in this respect. On the other hand, they had to be relatively tolerant against other religions (including Catholicism) for the sake of business. In practice, they seemed to have followed some kind of middle road (Stern 2011).

Political rights for the native population under the British East India Company ruling, was never really on the agenda (Marshall 1997).

Apart from these pieces of information, it has been difficult to find information specifically relating to the civil rights of local populations under company rule. The relation to the native population cannot be used straight off as an analogy with our scenario, however, since we here concern ourselves with the civil liberty of the settlers of space colonies. We assume that contrary to what was the case in earthly colonies, there will not be any native inhabitants on the planets or moons when they are first settled by earthlings (see Persson 2014). When trying to learn from history we should therefore focus on the civil liberties of the settlers rather than the natives.

When Virginia was ruled by the Virginia Company of London, the settlers apparently lacked both property rights and the right to a trial by jury (Nelson 2008). The circumstances in Virginia differed greatly compared with the neighbouring states that had other kinds of rule where they practiced English common law and had working courts and lawyers (Nelson 2008). It also changed in Virginia when the crown took over as governing power, apparently not in order to increase civil liberties but to reconstruct the economy of the state, though it actually did improve civil liberties at least concerning property rights and legal rights (Nelson 2008).

The Hudson's Bay territory did not have many settlers apart from the company employees. HBC was in fact opposed to outsiders settling in the area since they made the trapping more difficult by disturbing the animals and transferring forest to farmland (Newman 1985). It might be claimed that company employees were not really settlers since they only stayed in the territory for a limited time and then returned home. I choose to count them anyway, however, since this will probably also be the case in the first stages of space settlement.

In the Hudson Bay territory, the typical contract ran for 5 years, though most employees served more than one contract (Newman 1985; Payne 1989). That they choose to come back after their first contract could be a sign that they were generally happy with the way they were treated. Newman (1985) believes, however, that the explanation was that they had problems adapting to the life back home after having spent 5 years in the wilderness.

Company records regarding the employees tend to be less informative the lower the rank of the employees (Payne 1989). This mirrors the fact that the society, at least at the York factory, which was the chief factory for a long time, was organized in a very hierarchical and military style (Carlos and Lewis 2010; Payne 1989). Leitia Hargrave who was married to the head of the factory, James Hargrave, pronounced the York factory "the most respectable place in the territory" (Payne 1989). This exclamation had hardly anything to do with the state of civil liberties, though. Payne believes that it was the strict hierarchy and the far reaching efforts to arrange that the higher classes would not have to associate much with the lower classes that was behind her judgment (Payne 1989).

A strict hierarchy with strict boundaries between the classes was seen as important for upholding the discipline (Payne 1989). The company officers had unrestricted power over the lower ranks and the punishments were hard and arbitrary, though apparently usually not extreme (Payne 1989). 1792 it was decided that fines should be administered by the governor and council and that the London committee should review the decision (Payne 1989). This seems to be a step in the direction of a more humane and less arbitrary system.

Apparently the Hudson's Bay Company actively recruited employees that were used to a strict hierarchy and easy to control (Payne 1989). For that reason, employees from Scotland, especially the Orkneys, and Canada were preferred over those from England and Norway (Payne 1989). Even so, labour unrest was common at least at the factory in York (Payne 1989).

The company did not just exert a strict hierarchical power over its employees on the workplace. It apparently interfered in all aspects of the life of its employees (Newman 1985). Religion seems to have been an exception to this, however. There were attempts to get the employees to church but these attempts were not pushed particularly hard (Newman 1985; Payne 1989).

The company encouraged the employees to read, especially religious, technical and self-improving literature (Payne 1989). According to Payne (1989), officers and missionaries had a certain control of the book selections at the post libraries though I have not found any notes on outright censorship.

The strict social hierarchy that was fund in Canada was also prevalent among the British subjects in India (Marshall 1997), and just like with the employees of the HBC, most British employees of the British East India Company only stayed for a while and then moved back so they were not really settlers (Marshall 1997). Contrary to what was the case with the HBC, on the other hand, the British East India Company wanted to attract settlers (Stern 2011). One way of doing that was to establish laws and courts that could guarantee safety, property rights and due process for settlers (Stern 2011). They therefore, also contrary to the Hudson's Bay Company, had written laws and courts with appointed judges (Stern 2011).

The company also experimented with elected self-governments in some urban areas, though they were only allowed to govern themselves to the extent that it did not clash with the company's interests or undermined company authority (Stern 2011).

The British East India Company apparently tried to be more humane than the Dutch that had previously been in possession of some of the territories that was eventually taken over by the British. The latter also tried to find a balance between a system that facilitated maximum profit and a system that considered the wellbeing and safety of the civil society (Stern 2011). Even so, the justice for the settlers has been described as "imperfect, piecemeal, and inconsistent." (Stern 2011). There was also much discontent among settlers and soldiers (Stern 2011).

As we saw above, the company tried to find a middle way between religious freedom and the spread of Christianity among the locals. For the British in company territories "[s]ome degree of religious conformity to the established Church of England or Church of Scotland was … expected." (Marshall 1997). Exactly to what extent it was enforced seems difficult to assess, however. When it comes to freedom of expression and of political activity among the Europeans, in the company ruled territories in Asia, both were strictly repressed, just as it was for the natives (Marshall 1997).

Rebels both in Asia and in Virginia tended to claim that they would favour a colony under the crown (Stern 2011). This might indicate that company rule was considered worse than the rule of the crown, but it can just as well be a statement made for tactical reasons. In a conflict with the ruling company, it would have been very important for the rebels to maintain good relations with the crown.

In Northern Rhodesia, the relationship between the settlers and the British South Africa Company was notoriously bad (Slinn 1971). The company refused for instance to give the settlers permission to form an 'Advisory Council' (Slinn 1971), which would have given them some influence over the administration of the territory they inhabited.

A possible explanation for why the settlers were often less happy with corporate rule than the natives could be that settlers from Britain and other European countries were used to a higher degree of liberty from home. If this is at least part of the explanation, it indicates that future space settlers who are used to substantially more civil liberty than the people during the 17th, 18th and 19th centuries will not be less unhappy under corporate rule, unless the corporate government will grant its subjects a considerably higher degree of civil liberty than their historical counterparts did on the earth.

9.5 Some Philosophical Reflections

It is important to study and learn from history but it is also important to be aware of the limits of historical analogues. One potentially important difference between our case and the historical cases is that the historical cases resulted from a mercantilist system while the case we discuss is based on a libertarian system. This difference may have some important implications. One potentially important implication is that the historical cases dealt with monopolies, while there, at least in theory, may be more competition between corporations operating in space. This includes competition about the workforce as well as about customers. The competition about the workforce could result in a better bargaining position for the workers/settlers, which in turn could have positive effects on their civil liberties.

This difference should not be overstated, however. In many cases even the historical monopolies had competitors, especially in the form of corporations from other countries, but sometimes also from their own country. The Hudson's Bay Company had for a while a competitor in the form of the North West Company that they eventually merged with. They also competed with French companies (a competition that at some instances was settled by acts of war) and there were cases, though rare, of employees who escaped from one company to start working for another (Payne 1989). In Asia there was constant and fierce competition between different countries' East India Companies and also with so called interlopers. Also in these cases the competition and the double roles of the companies as traders and governments frequently led to armed conflict. It seemed to have been more common in Asia than in Canada that employees abandoned their employer for another employer or even for trading on their own (Stern 2011). In addition to the fact that many more people worked for East India Companies than for the Hudson's Bay Company and its competitors, the physical environment in northern Canada was an important explanation for why it was so uncommon that employees attempted to move between companies and even fewer did so successfully (Payne 1989).

The environmental conditions in space are even worse than in northern Canada, however. In space it will be necessary to have access to highly technological solutions that due to their technological complexity will be more expensive and more difficult to put together by oneself. This dependence on artificial life supporting systems will make it difficult to run away (Cockell 2013, 2015). It can also be expected that spacesuits and spacecrafts will be strictly controlled by the governing corporation. In fact, the most probable outcome of a situation where there is competition over workers and where the corporation and the government is the same entity, is that the corporation will enforce restrictions of the freedom of movement for their employees. To enforce such restrictions can also be expected to be much easier in a space environment where the corporation owns all available means of transportation compared to what is the case on the earth where more low-tech means of transportation can be sufficient for a successful escape.

Before making sure that people stay with the company, it will have to attract employees and possibly other settlers, however. As we saw in the previous section, the British East India Company tried to achieve that by improving the safety, property rights and due process to attract settlers. On the other hand, there will probably be no shortage of people on the earth who have nothing to lose and who are prepared to try their luck on another world almost independently of the circumstances on that world.

The competition over customers is maybe a more promising tool for improving the conditions of the settlers. This should be seen in the light of another difference between the historical analogues and a future space settlement governed by a corporation, namely that most earth societies are very different today than they were during the colonial era. Democracy and democratic attitudes as well as public concern over civil liberties are considerably more prevalent today. The social responsibility of companies is also much more highlighted and plays a much larger role for the reputation of a company than during the colonial era. In addition to all of this, news fly faster and are more easily spread with today's technology than they were then. On the other hand, if the corporation and the state are the same, it can be expected that it will be very difficult for journalists and independent organisations to get reliable information about the conditions for the workers/settlers on the world in question, which will make it difficult to use consumer power to put pressure on a corporation that does not respect the civil liberties of the settlers over whom it rules.

Another difference between historical earth colonies and future space colonies is that the space settlers would be absolutely dependent on the company in a way that has never been the case on the earth. Space is a very hostile environment even for those who do not try to run away. The very survival of the settlers will depend on highly technological and expensive systems. Breathable air is the most extreme example (Cockell 2013, 2015). We can survive without food for a few weeks but we cannot survive without air for more than a couple of minutes. If the organisation that governs the settlement is the same as the corporation that supplies the life supporting services, it will create an extreme asymmetry in power between the rulers and the ruled that will make it very difficult for the settlers to claim any respect for their civil liberties. It will also make it easier for the rulers to claim that it is necessary to maintain a very strict control and discipline since breaking the rules might endanger the whole settlement (Cockell 2013).

Even if this asymmetry in power is not actually wielded as a weapon in an actual conflict, its mere existence will have a psychological effect on the settlers. It will make them feel less free and it will also make it less likely that they take action against their leaders even in situations that are rather bad from a civil liberties point of view.

The power concentrated in an organisation that combines the roles of civil government and sole employer is already formidable. If this is also combined with the role of being the sole provider of oxygen and other life supporting services, the power of the corporation over the citizens will be much stronger than can be accepted in a liberal society, even if it is not actively used to oppress the citizens.

Some things will doubtlessly be the same whether we talk about historical earth colonies governed by corporations, or about future space settlements governed by corporations.

The raison d'etre for a corporation is to produce the greatest possible profit for its owners. This was made very clear to the local governor of Bombay when the committee in London very bluntly pointed out that "our business is to advantage ourselves by trade and what government we have is but the better to carry on and support that." (Stern 2011). It stood also clear when the colonial office in London showed sympathy with the settlers in Rhodesia, and the British South Africa Company answered that the colonial office lacked understanding of the commercial motives (Slinn 1971).

This will create a very serious tension between the corporation's responsibilities to its owners and its ability to govern the settlement according to liberal principles. The ultimate aim of laws, etc. decided by the company in this territory would have to be to produce the greatest possible profit for the owners. It would not be to guarantee the civil liberties of the inhabitants. Even if the local administrators have good intentions and really want to create a society with a high degree of respect for the civil liberties of the settlers, they will be restricted in their ability to live up to this ambition by their responsibilities to maximize profit for the owners. Even though civil liberties were prioritised by the British East India Company when it was beneficial for the company, as when they wanted to attract more settlers, it is obvious that the respect for civil liberties shifted between companies. The fact that the population was dependent on the sentiments of the company leaders and did not have any means of influencing the policy in a democratic way shows that civil liberties was not something one could count on as employee or settler in territories under company rule. The company strategy might imply respect for civil liberties or it might not, and the company board or at least individual decision makers in the company might have a positive attitude towards granting at least some civil liberties to at least some of its subjects, or they might not. Since the rulers did not answer to its subjects, even the most basic civil liberties were never something that could be counted on.

This bring us to the maybe most important thing space settlements governed by corporations would have in common with their historic counterparts on the earth, namely that the government would not be appointed in a general election among the settlers and it would not answer to those over which it rules. The government would be the board of the company which in turn is elected by the shareholders. Even if the settlers are given the right to appoint their own counsel, the ultimate power will still rest with the company board answering only to the owners.

The right to take part in the appointment of the leaders of one's society is a very basic civil liberty (Gearty 2007; Locke 1988) that includes the right to vote and all the rights that are necessary for this right to be meaningful, such as the freedom of thought and expression and the freedom to assemble (Gearty 2007). The absence of any of these basic rights will therefore be a very strong infringement of the civil liberties of the settlers.

In many cases, the rulers in the form of the board of the company would not even live in the territory over which they rule. They would, at least in the early days, typically reside on the earth. Extraterrestrial societies governed by a company would thus, perhaps ironically, still be ruled from the earth, just not by a political body in which the settlers might have some influence, but by the board of the company, just like when the "adventurers of England" were running the Hudson's Bay Company and thereby the Hudson's Bay territory. These "adventurers" had in fact with few exceptions never been in the Hudson's Bay area, but they still had unlimited power over the territory and its inhabitants (Newman 1985).

A specific problem that relates to all the complications pointed out above has to do with the right to due process and a fair trial. History has taught us that the individual or group "who possessed political and economic control of a colony routinely used their power to steer that colony's law in directions that would further their ends and goals." (Nelson 2008). In a situation where the settlers are absolutely dependent on the company for its survival, where the very reason for the company to exist on this world is to make money for its owners, and where there is no civil government appointed by and answering to the inhabitants that can limit the freedom of the company, the status of the courts will be very difficult. It will be very difficult for individual settlers to get a fair trial when accused by the corporation. It will also be very difficult to get the corporation convicted for crimes against the settlers. If the corporation refuses for instance the formation of unions, there would be no civil authority to which the workers could appeal.

To put it in more general terms, the specific circumstances we have found here would make the division of power between the government, the parliament and the courts that is a hallmark of a modern democracy, rather unlikely.

Finally, a corporation operating on a new world will be interested in a particular kind of settlers, that is, settlers who will be useful as workers in whatever enterprise the corporation is involved in on that world and at the same time do not cause any problems for the company. Given the overarching aim for the corporation of making money for the shareholders, "free thinkers" and dissidents will not be of any interest for the company. If the corporation also plays the role of government, they will have the liberty to choose who will be allowed to settle on the world and what educations will be offered there. This could have a very serious effect on the effective exercise of civil rights. The exercise of many civil rights requires that people have a basic knowledge of political theory, critical thinking and how to form and express ideas. It also requires the presence of people who are experts in critical thinking and in forming and analysing new ideas, including ideas about liberty and about government. Without these kinds of people, civil liberties are a dead concept. Des Rosiers (2012) argues that humanities in the form of for instance art, literature, history and philosophy, is essential for the exercise of civil liberties. This is probably true. It is probably also true for many other disciplines, but humanities might be the clearest example of an area that risks being excluded in an educational system that is formed by for instance a mining company on a foreign world. This is probably not true for all of humanities. Art that glorifies the company and its leaders will not just be accepted but actively promoted. Art that can work as a distraction for the people will probably also be encouraged. A company that operates on a foreign body will also need people proficient it different languishes, particularly if it is in the tourist business, but any kind of trade with the earth on a global level will need some language proficiency. This means that languages will probably also be taught at schools and universities under company ruling. Other, more 'dangerous' areas, such as philosophy, on the other hand might well be excluded or at least severely restricted so that only 'harmless' ideas are taught in the schools and at the universities.

9.6 Conclusion and Recommendations

What we have seen above leads us to the conclusion that the civil liberties of individual citizens might not be a high priority in a corporate state.

This means that even if extended economic liberty in space is a powerful incentive for the establishment of human settlements in space, and even if the right for corporations to rule over settlements on worlds where they operate would strengthen these incentives and in many ways be a convenient solution that will help the settlement to stay independent form earth governments, it will not be beneficial for the civil liberty of future settlers.

Based on this conclusion, I therefore suggest that any increase in economic liberty in space should not include the right for corporations to rule over space settlements. I also suggest that the international community on the earth today takes the question of extraterrestrial liberty seriously and starts working out a plan for securing that the power over any extraterrestrial society will be in the hands of the people from day 1 even if the settlements are established by, or otherwise the result of the activities of a corporation.

This is not a recommendation not to allow commercial interests in space. Neither is it a recommendation not to accept or make use of commercial driving forces for establishing space settlements. What it means is that the power over the civil society, including the power to make and enforce laws, must be in the hands of the people also in space and kept separate from any commercial interests. It also means that efforts to increase economic liberty in space will have to stop short of allowing corporations to rule over space settlements and be complemented with appropriate restrictions to guarantee the civil liberty for future settlers.

This kind of restrictions of economic liberty might sound like a rather anti-libertarian suggestion, but I am convinced it will lead to more substantial civil liberty for the individual settlers in the end. In fact, one of the most basic principles of libertarianism is that it does not accept that the liberty of one group is used to limit the liberty of others (Locke 1988; Mill 1983). This means that the economic liberty of corporations cannot imply that the corporation is at liberty to restrict the civil liberty of the settlers including their liberty to appoint their own government and all liberties that are necessary to provide substance to that liberty.

References

Beery, J. (2012). State, capital and spaceships: A terrestrial geography of space tourism. *Geoforum, 43*, 25–34.

Giacalone, J. A. (2008). Global trends in the commercialization of space. *Journal of Business and Economics Research, 6*, 65–76.

Beavin, M. (2008). Office of space commercialization promotes U.S. space industry. *Earth System Monitor, 16*, 7.

Carlos, A. M., & Lewis, F. D. (2010). *Commerce by a frozen sea*. Philadelphia: University of Pennsylvania Press.

Cockell, C. S. (2013). *Extra-terrestrial liberty—an inquiry into the nature and causes of tyrannical government beyond the earth*. Edinburg: Shoving Leopard.

Cockell, C. S. (2015). Freedom in a box: Paradoxes in the structure of extraterrestrial liberty. In C. S. Cockell (Ed.), *The meaning of liberty beyond earth*. London: Springer.

Bush, L. (2002). International space station commercialization. *Policy Technology in Society, 24*, 69–75.

Collins, D. (2008). Efficient allocation of real property rights on the planet mars. *Boston University Journal of Science and Technology Law, 14*, 201–219.

Collins, P. Q., & Ashford, D. M. (1988). Potential economic implications of the development of space tourism. *Acta Astronautica, 17*, 421–431.

Deep Space Industries. Retrieved June 22, 2014, from http://www.deepspaceindustries.com.

Des Rosiers, N. (2012). Civil liberties need humanities. *English Studies in Canada, 38*, 9–19.

DFI International 2002. *Market opportunities in space: The near term roadmap*. U.S. Department of Commerce, Office of Space Commercialization.

Dobkin, D. S. (1976). The supreme court, civil liberties, and the canadian bill of rights. In S. L. Wasby (Ed.), *Civil liberties* (pp. 191–199). London: Southern Illinois University Press.

Ehrenfreund, P., Race, M., & Labdon, D. (2013). Responsible space exploration and use: Balancing stakeholder interests. *New Space, 1*, 60–72.

Federal Aviation Administration. (2014). *Office of commercial space transportation*. Retrieved June 22, 2014, from http://www.faa.gov/about/office_org/headquarters_offices/ast/.

Gearty, C. (2007) *Civil liberties*. Oxford: Oxford University Press.
Gren, E. (1969). *Indiens historia*. Stockholm: Natur och Kultur.
Hertzfeld, H. R. (2002). *Space economic data*. Washington: The U.S. Department of Commerce, Office of Space Commercialization.
Hubbard, S., Davidian, K., Isakowitz, S., Logsdon, J., McMurry, J. R., Nield, G., et al. (2013). Growing the future of commercial space. *New Space, 1*, 3–9.
Lewicki, C., Diamandis, P., Anderson, E., Voorhees, C., & Mycroft, F. (2013). Planetary resources —the asteroid mining company. *New Space, 1*, 105–108.
Locke, J. (1988). *Two treatises of government*. In P. Laslett (Ed.). Cambridge: Cambridge University Press.
Marshall, P. J. (1997). British Society in India under the East India Company. *Modern Asian Studies, 31*, 89–108.
Meyer, Z. (2010). Private commercialization of space in an international regime: A proposal for a space district. *Northwestern Journal of International Law and Business, 30*, 241–261.
Mill, J. S. (1983). *On liberty*. New York: Penguin books.
Moon Express. (2014). Retrieved from http://www.moonexpress.com/.
Murthi, K. R. S., Sankar, U., & Madhusudhan, H. N. (2007). Organizational systems, commercialization and cost-benefit analysis of Indian space programme. *Current Science, 93*, 1812–1822.
NASA HQ Library. (2014). *Space commercialization and space tourism*. Retrieved June 22, 2014, from http://www.hq.nasa.gov/office/hqlibrary/pathfinders/spacetourism.htm.
NASA Policy Directive 7410.3 J.
NASA Policy Directive 8610.12G.
Nelson, W. E. (2008). *The common law of Colonial America: Volume I: The Chesapeake and New England 1607–1660*. Oxford Scholarship Online.
Newman, P. C. (1985). *Company of adventurers* (Vol. I). Markham: Penguin Books Canada.
Newman, P. C. (1989). *Empire of the Bay: An Illustrated History of the Hudson Bay Company*. New York: Viking Books.
Payne, M. (1989). *The most respectable place in the territory—everyday life in Hudson's Bay Company Service York Factory, 1788 to 1870*. Ottawa: Studies in Archaeology Architecture and History, National Historic Parks and Sites Canadian Parks Service Environment Canada.
Peeters, W. (2002). Effects of commercialisation in the European space sector. *Space Policy, 18*, 199–204.
Peeters, W. (2003). Space commercialization trends and consequences for the workforce. *Acta Astronautica, 53*, 833–840.
Persson, E. (2014). What does it take to establish that a world is uninhabited prior to exploitation? A question of ethics as well as science. *Challenges, 5*, 224–238.
Planetary Resources. (2014). Retrieved from http://www.planetaryresources.com/.
Reinstein, E. J. (1999). Owning outer space. *Northwestern Journal of International Law and Business, 20*, 59–97.
Schackleton Energy. (2014). Retrieved from http://www.shackletonenergy.com/.
Shi, T., & Lou, D. (2010). Subjective evaluation of changes in civil liberties and political rights in China. *Journal of Contemporary China, 19*, 175–199.
Skaaning, S.-E. (2009). Measuring civil liberty: An assessment of standards-based data sets. *Revista de Ciencia Política, 29*, 721–740.
Simberg, R. (2012). Property rights in space. *The New Atlantis, 37*, 20–31.
Slinn, P. (1971). Commercial concessions and politics during the colonial period: The role of the British South Africa Company in Northern Rhodesia 1890–1964. *African Affairs, 70*, 365–384.
Spaceport America. (2014). Retrieved June 22, 2014, from http://spaceportamerica.com/.
Spaceport Sweden. (2014). Retrieved June 22, 2014, from http://www.spaceportsweden.com/.
Stern, P. J. (2011). *The company-state—corporate sovereignty & the early modern foundations of the British Empire in India*. Oxford: Oxford University Press.
Sturm, A. L., & Wright, K. M. (1976). Civil liberties in revised state constitutions. In S. L. Wasby (Ed.), *Civil liberties* (pp. 179–189). London: Southern Illinois University Press.

Tunick, M. (2006). Tolerant imperialism: John Stuart Mill's defense of British rule in India. *The Review of Politics, 68*, 586–611.
UN. (1966). *International covenant on civil and political rights*. The General Assembly of the United Nations.
U.S. Department of Commerce. (2014). *Space commerce*. Retrieved June 22, 2014, from http://www.space.commerce.gov.
Velocci, A. L., Jr. (2012). Commercialization in space—changing boundaries and future promises. *Harvard International Review, 4*, 49–53.
Wasby, S. L. (1976). Introduction: Civil liberties and policy making. In S. L. Wasby (Ed.), *Civil liberties* (pp. ix–xiv). London: Southern Illinois University Press.

Chapter 10
Providing for Sustainable Exploration and Use of Outer Space Environments

John D. Rummel

Abstract Due to the policies and investments made by spacefaring nations and international space agencies over the last five decades, non-State Parties are now developing significant capabilities in space exploration and use. This is essential if there is to be the development of a true space economy. To ensure that these capabilities can lead to an economy that is both beneficial and sustainable, legal and regulatory regimes associated with the use of other planetary environments must anticipate them, and the roles that can be played by non-State Parties in the future development of space. While non-State Parties launching from a State signatory to the 1967 Outer Space Treaty (OST) are subject to the Treaty's articles, there are many shortcomings of the OST with respect to future space development. For example, it is clear that different launching venues could differ greatly in their implementation and enforcement of the Treaty provisions, inviting abuses of those provisions if a State's enforcement is ineffective. The possible short-term economic advantages of a "soft" interpretation of OST provisions may lead to conflicting uses of outer space resources, while a blind eye to the potential for "harmful contamination" of other planetary environments or the Earth could destroy the benefits of space exploration. It is thus timely to consider steps by which it would be possible to clarify and complement the OST's legal regime for the exploration of the Moon and other celestial bodies. There appear to be several approaches for extending the provisions of the OST and developing an international framework to accomplish these goals while providing for both protection and use of outer space as part of a future era of exploration, commercialization, and human habitation.

Keywords International law · Outer space treaty · United nations · Legal systems · Liberty

J.D. Rummel (✉)
Department of Biology, East Carolina University, Greenville, NC 27858, USA
e-mail: rummelj@ecu.edu

J.D. Rummel
Institute of Air and Space Law, McGill University, Montreal, Canada

C.S. Cockell (ed.), *Human Governance Beyond Earth*,
Space and Society, DOI 10.1007/978-3-319-18063-2_10

10.1 Space Commercialization and Its Implications

The five decades and more of investments made by spacefaring nations in basic spaceflight capabilities, coupled with recent policy changes by international space agencies, have fueled a wide spectrum of interest by commercial entities in space flight and exploration. For example, NASA's "commercial crew" initiative for the International Space Station (cf., Stone et al. 2008) was a specific move to develop commercial spaceflight for cargo and personnel support, while the *Global Exploration Roadmap* of the International Space Exploration Coordination Group (2013), a forum supported by 12 space agencies worldwide, lists as "Common Goals and Objectives" the mandate to "Stimulate Economic Expansion: Support or encourage provision of technology, systems, hardware, and services from commercial entities and create new markets based on space activities that will return economic, technological, and quality-of-life benefits to all humankind" (p. 3). Elsewhere, corporations such as Bigelow and Virgin Galactic are working to provide access to space for tourists and to supplement the capabilities of space agencies in accomplishing their scientific and exploration objectives. Likewise, companies like Shimizu (in Japan), and SpaceX (in the US) have espoused their own space agendas that have included hotels on the lunar surface and commercial expeditions to Mars and beyond. Perhaps even more of interest in the mid-term are "space mining" companies that have sprung up, including Planetary Resources, Shackleton Energy Company, and Deep Space Industries. While these would-be resource acquisition companies hope to find precious materials in space environments for their use and resale, a harder job, initially, will be to find an entity (*any* entity!) which could give them legal permission to do so.

These commercial space activities and concepts represent only the beginning of commercial space. If there are appropriate steps taken both to incentivize commercial activity (such as that provided in the ISECG Roadmap), and to remove disincentives that prevent investment into such capabilities by the commercial sector (such as provisions stating that space and its "exploration and use… shall be the province of all mankind and shall be carried out for the benefit and in the interests of all countries" (Moon Treaty 1979, Article 4). Future space activities by the private sector will require an understandable and appropriate regulatory regime if they are to make any progress.

The development of a commercial space sector in trans-lunar space and beyond is highly desirable. Such development is essential if there is to be a true "space economy," providing for eventual human movement off the Earth and the potential to develop a home for humanity that spans the rest of the solar system. Given the possibility of future large impact events on Earth (asteroids, comets), it may even be said that the future of our species is tied to the success of commercial space.

Tempering such a supportive view, however, is the realization that the effects of the profit motive on space environments will need some regulation. Not all of the profits to be made in outer space are commercial, nor even material/monetary, so the ability to protect outer space environments for science, and to protect the Earth overall, will be essential. Currently, both COSPAR and the International Academy of Astronautics (Rummel et al. 2012; Ehrenfreund et al. 2013; Hofmann et al. 2010) have called for specific measures to be taken to protect those outer space environments needing it (and by omission, opening up other environments suitable for commercial use) and the development of management guidelines and regulations for environmental protection to supplement the accepted regulations for preventing harmful planetary contamination of a biological or organic-chemical nature (e.g., COSPAR 2011).

A balance between the preservation of the beneficial aspects of space environments and the rise and successful integration of profitable, non-state actors in outer space can only be drawn successfully if the legal and regulatory regimes associated with the use of other planetary environments anticipate that rise. For example, in the recent International Academy of Astronautics report on "Protecting the Environment of Celestial Bodies" (Hofmann et al. 2010, p. 17) it was noted that "protecting the space environment is a significant and long-term task, but this does not mean that time is on our side. Once commercial endeavors begin, they can evolve very quickly." That quickness is both a blessing and curse—the latter if the regulatory framework is not prepared to deal with it.

As to specific concerns regarding commercial development of planetary environments, the PECB report includes the following (p. 49):

> When exploitation of celestial bodies occurs, major problems of contamination and environmental degradation will doubtless emerge. Terrestrial experience indicates that mining or other recovery of natural resources is contaminative of surrounding areas and damaging to the natural environment. If manufacturing or processing is carried out in situ, industrial processes would be involved with the import of materials foreign to the celestial body, and the creation of unnatural waste might occur. Suitable licensing and enforcement regimes will be required.

10.2 Establishing the Required Legal Regime(s): What Are the Choices?

Suitable licensing and enforcement regimes for the protection and use of outer space environments do not currently exist. These will have to be developed carefully and conscientiously with the goal of their continuing development in mind. In seeking a comprehensive regulatory regime for the protection of planetary environments, the same regulations will, of necessity, identify environments that are not subject to protection and allow for their use, whether for commercial development or for some other purpose. Such a regime could also offer the framework for a licensing solution that would define practically what "for the benefit and in the interests of all countries" would mean in practice.

The principles necessary for the establishment of a regulatory regime to protect planetary environments already exists in the Outer Space Treaty of 1967 (OST 1967).

The OST provides for protection against "harmful contamination" (Article IX) as an important consideration, but even more importantly, it contains the Article I provisions that consider the use of space environments, cf.:

> The exploration and use of outer space, including the Moon and other celestial bodies, shall be carried out for the benefit and in the interests of all countries…and shall be the province of all mankind. Outer space, including the Moon and other celestial bodies, shall be free for exploration and use by all States without discrimination of any kind.…

With the OST as a basis for a future regulatory regime, one can anticipate that outer space can be used by a wide range of interests and both direct and indirect benefits accrued to society. The OST, however, provides for almost none of the mechanisms that would be required to regulate those interests. In order to craft an effective instrument for governance, providing for both the protection of outer space and its use in a time of expanding exploration capabilities by private, commercial, entities, then it is clear that the OST will have be supplemented by at least one additional agreement. There are several problems inherent with the current treaty that need to be taken into account when providing for the future.

1. Finding a Sovereign: Who Owns the Solar System? The framers of the OST envisioned an international agreement that would be embraced by the major space powers (chiefly the US and the USSR) at the time, and therefore complied with as a matter of cold-war political necessity. The OST did not divide up outer space between those countries, but declared (OST 1967; Article I) that outer space is "the province of all mankind." This provision, along with the prohibition (Article II) against "national appropriation by claim of sovereignty, by means of use or occupation, or by any other means" is very clear, so far as it goes. By the OST, everybody owns outer space, and thus, from a practical perspective, nobody owns outer space. That has seemed just fine up until now, but if one needs to get the permission of the owner to do something out in the solar system (mine it, protect it, develop a hotel, etc.) then currently one is out of luck. As a consequence, any future elaboration of the principles of the OST must involve at least one mechanism that is inherently international.

The current signatories of the OST (the "States Parties" as they are known) will have to agree that they, collectively, can and should provide for the transfer of rights, at some level, from them as framers of the OST, to others. One can envision that a convention developed by the States Parties could result in an "Interplanetary Authority" (ILA) of some sort that would be empowered to license solar system locations for the use by other entities, whether they are individual nations or individuals or multi-national corporations. There is a similar entity that was established as part of the UN Convention on the Law of the Sea (the "International Seabed Authority;" cf., UNCLOS 1994, Part XI, Sect. 4), although there are a number of provisions about the seabed authority in UNCLOS that would not bear repeating in outer space (e.g., Malone 1984). In passing, it is useful to note that the extension of the United Nations OST by a convention does not necessarily require that the resulting convention is overseen through the United Nations, itself.

2. Who Gets to Play? Under the current OST (Article VI), non-State Parties launching from a State signatory to the 1967 Outer Space Treaty are subject to the Treaty's articles. Article VI also requires that "The activities of non-governmental entities in outer space, including the Moon and other celestial bodies, shall require authorization and continuing supervision by the appropriate State Party to the Treaty." It is clear that different launching venues can differ greatly in their implementation and enforcement of the Treaty provisions (COPUOS 2001), inviting abuses of those provisions if a State's enforcement is ineffective or incomplete. As stated in the PECB report:

> The…assumption is that there is no current consensus or "normative understanding" among those states with space capabilities. These states may have their own standards for protecting celestial bodies, but that means that there are different interpretations… If one of these standards is not fully protecting celestial bodies, it would undermine all the efforts that other states may take (Hofmann et al. 2010, p. 66).

Under the OST, the assumption is that the States Parties are the responsible entities, and that actions under the OST are their responsibility. Only they "play" in the arena of the OST, and thus allocations of space environments might be expected to only go to them, and they might be expected to provide for the "supervision" to protect environments in space. Nonetheless, if we can posit a regime where the allocation of space assets will be governed by an international entity that would be subject to the assent of the States Parties to the convention, then it would also be possible for the convention to allow for licensure and other interactions at a level below that of the States Parties themselves. When developing specific measures to be taken to protect outer space environments needing such protection, it may be essential for the "International Licensing Authority" to be able to deal directly with individuals and corporations, and not just with the current governments represented in the group of the States Parties.

3. A Thing of Beauty is a Joy Forever? In consideration of the time and effort needed to undertake commercial activities in outer space, the timespan and renewal procedures associated with the allocation and licensing of extraterrestrial environments—as well as their protection, where warranted—will be critical issues in the development of an ILA. The nature of the protection/use planned for different environments will also be a factor in making timespan determinations.

4. Buying or Renting? If human commerce and capabilities are extended into the solar system, the issue of private property will surely follow. At one level, it will harder to enact if an ILA allocates only limited rights to a national government representing one of the States Parties, which may not be motivated to secure private property rights for its citizens outside of its sovereign territory. At the other end of the scale, a conscious decision by the ILA to provide a route to private ownership for individuals might be made as a means of dealing with political pressures brought to bear by corporations or individuals with a long-term and intensive association with specific outer space environments. Practically speaking, providing a pathway to legitimize what it cannot stop might be an attractive option for the ILA as it develops.

5. A Rising Tide Lifts All Boats? One of the most criticized aspects of the UNCLOS was the nature of the provisions intended to provide for the sharing of benefits from seabed mining and other activities. Not only could the International Seabed Authority levy fees for licenses allocated, but they could use the licensee's intellectual property to compete with them in the same business, on behalf of "all mankind." A less detailed but similarly intended provision is given in the Moon Treaty (Article 11 (7)(d)) which provides for "an equitable sharing by all States Parties in the benefits derived from [lunar] resources, whereby the interests and needs of the developing countries, as well as the efforts of those countries which have contributed either directly or indirectly to the exploration of the Moon, shall be given special consideration." The use of a portion of the fees collected by the ILA for any allocated outer space license in order to provide for States Parties trying to develop their capabilities in space would be an appropriate response to benefit-sharing intentions. Any such fee sharing could eventually provide for a more capable and diverse human presence in outer space, without disincentivizing those who are first ready to undertake commercial risks in outer space.

10.3 Requirements for a Licensing and Enforcement Regime

The development of an framework to protect extraterrestrial environments within an agreed-upon international strategy for the exploration, commercialization, and human habitation of space is necessary to provide for the needs of future space enterprise and human habitation off of the Earth. One vehicle for such a framework could be developed by a convention established under the OST, which would establish a specific entity (the ILA) whose job would be to provide uniform, internationally enforceable, provisions for the use and protection of outer space environments by any other entity, whether an individual, a corporation, a government, or another intergovernmental organization.

The ILA would be the means by which the States Parties to the OST monitor and implement the provisions of the treaty in the modern era of space exploration and use, which will include both state-and non-state actors, and may embrace a variety of goals and objectives ranging from purely personal (vanity tourism) to rigorously scientific (e.g., Europa surface landers) to purely commercial (selling ice-mined water to bases on the Earth's Moon). The gamut is wide, and many such enterprises will seek to be part of humanity's activities in outer space. For example, while it is likely that fishing and logging will never be part of space commerce, but it is almost certain that aquaculture and farming will be.

In order to establish such an entity and operate it successfully, the following organizational capabilities should be considered for inclusion into the ILA:

- A capability for identifying important extraterrestrial environments and monitoring their preservation and/or development.

- A planning process for the sustainable use of extraterrestrial resources, including those which are unique environments/opportunities for scientific study.
- Mechanisms for continuing technical and scientific cooperation among the states parties.
- A capability to sponsor research on extraterrestrial environments and provide for the training of future experts.
- A strong program of public education and awareness.
- A mechanism for public participation in assessing the environmental impacts of development projects and planetary surface exploration.
- A mechanism to foster the free exchange of publicly available information.
- Expertise in the valuation of commercial activities in space to understand the balance between risk, investment, and potential profitability.
- A straightforward and open licensure process for the use of outer space environments, to include steps to be taken to establish private property rights in outer space.
- An appropriate tribunal to adjudicate disputes among the parties.
- Legal expertise and an enforcement arm (including internal affairs checks and balances).
- The ability to call on States Parties for assistance in enforcing the ILA's provisions of licensure.
- The capability to use licensure fees to provide for the maintenance of the enforcement regime (e.g., the ILA, itself, and any other capabilities needed).
- The capability to use licensure fees to provide for any "common heritage of mankind" provision that may be seen as desirable by the establishing convention.

The development of the ILA cannot be accomplished overnight, but in whole or in part will be necessary to the development of future space commercialization beyond low-Earth orbit.

10.4 Next Steps

Non-governmental organizations such as COSPAR (the ICSU Committee on Space Research) which represents national science institutions from 44 member countries, 13 international scientific unions, and 10 associated companies, can incubate the formation of a convention of the States Parties to the OST, and presumably would eventually work on such a convention with the United Nations Committee on the Peaceful Uses of Outer Space (COPUOS). COSPAR strives to promote the use of space science for the benefit of mankind and for its adoption by developing countries and new space-faring nations. Scientific interests are paramount to COSPAR, and essential to successful human capabilities expansion.

Common regulations need to be elaborated to ensure valuable, safe, economic, and broadly-based exploration that encompass and balance a diverse set of stakeholder interests and will benefit both current and future generations. To protect science and the future, there needs to be a focus on protection, where it is needed, involving both more and less stringent requirements than are now in effect. This would include the accepted regulations for preventing harmful planetary contamination, which currently only consider biological and organic chemical contaminations.

To aid in the development of a convention under the OST, COSPAR could help elaborate management guidelines and draft regulations in interaction with other organizations such as IISL, UN-COPUOS and others. A first step would be to establish a framework for environmental stewardship, embodying existing environmental regulations into a framework that can be shaped into the ILA in the future. Meanwhile, a study of some of the questions raised earlier, and their implications to space commercialization, can proceed as well.

10.5 Summary and Conclusion

Non-State Parties are developing significant capabilities in space exploration and use. Such entities are governed by the 1967 OST, but enforcement is not specified there; the different launching venues that are responsible may differ greatly in their implementation of the Treaty provisions. In addition, no entity under the OST is charged with promoting either protection or use of outer space environments. As such, it is becoming necessary to clarify, complement, and buttress the legal regime currently regulating the exploration of the Moon and other celestial bodies. Individual states/parties cannot regulate these activities individually, and consensus on an approach and mechanisms is required to embody the principles of the OST into an operating entity that can provide for future human expansion into outer space. Some ideas have been presented on what that operating entity should comprise, and what activities it should envision in working with governments, corporations, and individuals in the future allocation of outer space environments.

References

COPUOS (Committee on the Peaceful Uses of Outer Space). (2001). Review of existing national space legislation illustrating how States are implementing, as appropriate, their responsibilities to authorize and provide continuing supervision of non-governmental entities in outer space, UN Doc. A/AC.105/C.2/L.224

COSPAR. (2011). Planetary Protection Policy (revised 24 March 2011). *COSPAR, Paris, France.* Retrieved form https://132.149.196.38/sites/default/files/pppolicy.pdf.

Ehrenfreund, P., Herzfeld, H., & Howells, K. (2013). *Workshop Report: COSPAR workshop on developing a responsible environmental regime for celestial bodies.* Space Policy Institute, GWU, and COSPAR, Washington, DC. Retrieved from http://www.gwu.edu/~spi/assets/docs/PEX_WorkshopReport_ES_March14_Web%20%281%29.pdf.

Hofmann, M., Rettberg, P., & Williamson, M. (Eds.). (2010). *Protecting the environment of celestial bodies (PECB)*. Paris, France:IAA Cosmic Study, IAA.
Malone, J. L. (1984). Who needs the sea treaty? *Foreign Policy, 54*, 44–63.
The International Space Exploration Coordination Group. (2013). *The global exploration roadmap (2nd release)*. Retrieved from www.globalspaceexploration.org.
Moon Treaty. (1979). Agreement governing the Activities of States on the Moon and Other Celestial Bodies, adopted by the General Assembly in its resolution 34/68, opened for signature on 18 December 1979, entered into force on 11 July 1984.
Outer Space Treaty [OST]. (1967). Treaty on Principles Governing the Activities of States in the Exploration and Use of Outer Space, including the Moon and Other Celestial Bodies, adopted by the General Assembly in its resolution 2222 (XXI), opened for signature on 27 January 1967, entered into force on 10 October 1967.
Rummel, J. D., Race, M. S., Horneck, G., & The Princeton Workshop Participants. (2012). Ethical considerations for planetary protection in space exploration: A workshop. *Astrobiology, 12*, 1017–1023.
Stone, D., Lindenmoyer, A., French, G., Musk, E., Gump, D., Kathuria, C., et al. (2008). NASA's approach to commercial cargo and crew transportation. *Acta Astronautica, 63*, 192–197.
United Nations Convention on the Law of the Sea. (1994). Opened for signature December 10, 1982, entered into force 16 November 1994. 1833 U.N.T.S. 3

Chapter 11
Constituting Outer Space: The Governance of Planetary Settlements and Artificial Habitats

Mukesh Bhatt

Abstract Activities, habitats and settlements in space will inevitably be subject to human governance. The content of a manifesto, 2 proposed constitutions, and proposed US legislation for habitation and exploitation off Earth are presented and examined from a generic constitutionalist perspective, within a proposed framework for space governance based on the US model. This concerns the locus of authority, involves the rights and responsibilities of the general populace and the structure of government, with particular reference to the concepts of countervailance, of sovereignty and of jurisdiction, considered with reference to the peculiar characteristics of the spatial environment. The analysis leads, among other conclusions, to several possible solutions for establishing autonomous sovereign states in space with respect to international law, and shows the persistence of Earth based values and ideologies.

Keywords Constitutionalism · Governance · Sovereignty · Jurisdiction · Space habitats · Planetary settlements

11.1 Introduction

Humans are at the point of being able to direct their own evolution. As the exploration and exploitation of space progresses, humans will move to live in space. Such habitats may be established at LaGrange points, as space stations, near Earth or on planets, moons and asteroids, or exist as interstellar and interplanetary generational spaceships travelling through the void.

In space, all habitats beyond Earth may be considered to be inimical to human life, especially as humans are bio-fragile (Robinson 2006). Habitats in space have their own added necessities and requirements, being dangerous in their own right,

M. Bhatt (✉)
School of Law, Birkbeck College, University of London, London, UK
e-mail: m.bhatt@physics.org

C.S. Cockell (ed.), *Human Governance Beyond Earth*,
Space and Society, DOI 10.1007/978-3-319-18063-2_11

where any individual activity can affect the collective safety of the general population. Such habitats, as seen in the International Space Station, may involve and contain multiple cultures with varying and differing values and beliefs, but all will share a need to obtain essential resources from outside sources at all times, despite a presumed goal of self-sufficiency. The hostile environments found off-Earth, and some on Earth demand an equitable partition of common necessities, food, water, and air at the least. All terrestrial activities and imperatives (communality, security, environment, human agency, crime etc.) are magnified and intensified in space because of the lack of readily available resources.

Human habitats, whether on Earth, or established off it, all require a form of administration according to some set of rules, the need for such regulation having existed for the millennia and more that human societies have done. Recently constitutions have been proposed in high school projects for settlements on Mars (for example see Merritt 2010; Kelly 2012; Metcalfe 2012), especially since Kim Stanley Robinson's Mars novels. Schools in Pasadena and Washington State in the USA cover constitutional governance in space. The Virginia Aerospace Science and Technology Scholars program *Mars Live!* (*Mars Live!* found at http://vasts.spacegrant.org/modules/10) guides students through the basics of setting up a colony, with its module 8 dedicated to colony governance; this is a residential 7-day course sponsored by NASA and the Virginia Space Grant Consortium for high school pupils in Virginia. The inclusion in the K12 educational curriculum helps propagandize and maintain interest amongst a newer generation of space enthusiasts; compare the use of cigarette advertising or of cereal boxes through the decades in the US, Russia and China to advertise and encourage this response, or the 2014 request and competition in Britain to suggest and design dinners for the ISS. Earlier suggestions for governance date back to the 70s and 80s, several manifestos and constitutions appearing as part of general culture then, making these some of the earliest now available online. However, Burgess (1987), *Libra* (World Research Inc. 1978), the Minervan constitution tentatively dated 1977 (*Starship Aurora—Constitution of the Republic of Minerva* 2001-2) and the debates of the 80s workshops (Robinson 1988, 2004, 2006) are representative of the early enthusiasm of the first period of the space age.

Given the contextual introduction above, there follows a discussion of the study of constitutions and constitutionalism outlining current approaches to its central issues. Four samples clearly limited in style and approach to US models of governance in space are then presented and discussed in context, these being selected primarily on the basis of availability and age, K12 examples being excluded for reasons of brevity.

Libra was a privately funded, self-governing space station proposed in 1978 (World Research Inc. 1978). The "first" Space Age also provides a context for the so-called Tharsis Montes Commonwealth (Burgess 1987) on Mars, with its constitutional guarantee of civil rights based on a procedural approach. An economic and regulatory constitution dating possibly from 1977 protects personal rights whilst refusing personal responsibilities for the capitalist Republic of Minerva on the Starship Aurora (*Starship Aurora* 2001-2), and is proposed by an Ayn Rand inspired Objectivist society. Finally, in 1981 a Bill presented by Gingrich and others to the US House of Representatives (Gingrich 1981) can be viewed as a first step towards the exploitation of mineral resources, and safeguarding constitutional protections whilst providing a route-map for eventual sovereign statehood. Apart from content, ancillary information from accompanying documents will be discussed for points and issues specific to the situation. All relevant ancillary documentation is available online, accessible from their respective home pages.

11.1.1 Constitutions and Constitutionalism

All human societies have sets of rules and regulations that guide, limit and allow types of behaviour, and which allows for the peaceful or warlike co-existence of individuals within a given society. Hammurabi's code dates back to c. 1754 BCE and was inscribed in stone, as were the edicts of the emperor Aśoka (269-232 BCE) of the Mauryan empire in the Indian sub-continent, which were added to in later centuries by the rulers of other empires. The Licchavi republic, so-designated by scholars, apparently operated under an oral constitution (Mukerji 1969) and is described in the extant Buddhist literature as dating back to c.600 BCE. Codes of Law were also compiled in the Indian sub-continent, orally from at least 1200 BCE, and written down from about 800 BCE, the best known of these being the social and procedural codes of Manu and of Yājñavalkya, still in use to the present day amongst Hindu communities, and the *Arthaśāstra* by Kauṭilya, from about 300 BCE, which describes the form and functions of government and the rights and responsibilities of persons.

Various codes were also compiled in the Roman empire, the best known and most recent being the *Corpus Juris Civilis*, or Code of Justinian issued between 529 and 534 CE, and is the basis of much of the civil codes of Europe and the later constitutions of France and the USA. More recent examples of written constitutions include the Weimar Republic in the 20th century, the purely defensive constitution imposed by the Americans and other Western powers on Japan after the second World War, and that of Mongolia. The most well known document often seen as a pre-cursor or substitute to constitutions in common law jurisdictions is the *Magna Carta Libertatum* signed in 1215 CE, which limited the powers of the English monarch and devolved rights and privileges to the church and a group of rebel barons. The United Kingdom, although often referred to as a constitutional monarchy, does not have a constitution, as also some other countries. National constitutions since 1789 have had an average life of 17 years (Ginsburg 2007).

Stating what a constitution is not easy, as it immediately leads into a discussion of various issues concerning not just definitions, but content and domains, whether legal, political or otherwise. A working model descriptive of what the present author sees as constitutive of a constitution is presented below, and will be expanded upon where necessary. However, the discussion as presented is not intended to be comprehensive or complete, and the interested reader is directed to the vast literature on the subject.

First described and accepted orally, and generally passed down through the generations in the form of stories or exemplary behaviour, rules and regulations that reflect values governing behaviour between individuals, administrations and outside societies would have been enshrined in codes of law, given the later rise of authority structures. The first general and possibly most central problem of constitutional study is thus the origin, source, and right, privilege or prerogative by which such laws are promulgated. Note here the distinction between codes containing laws and an actual written constitution. Although much overlap exists, a law is often specific to a given situation, whereas a constitution is likely a more general statement from which such laws may be derived. Thus, a constitution supposedly lays out in quasi-religious terms the values of the collectivity: what should be, rather than what is, regarding the governance and behaviour of the population. In short, a constitution is a document that states, often in a written form, the rights, duties, responsibilities, privileges, obligations and prerogatives where these exist of the individual elements in a society, whether these be persons, institutions or governing authorities at the least.

The constitution also defines persons, institutions and authorities within the society, their form and structure, and the relationship between these separate persons, institutions and authorities; it lays down processes and procedures for the formation and operation of these institutions and authorities, and in a sense, by defining who belongs to the social polity, forms the person in that society. The constitution further limits or extends the powers of individuals, institutions and authorities whilst defining relationships between authorities, institutions and individuals. Given that persons are multiple entities, and where there are several institutions or several authorities, the processes also extend to relationships between these and at each level, that of groups, separate social polities, and the like.

In the modern world, all social and political polities are considered to require the limitation of authority in some way, and this is generally done by the use of a constitution. That a nation-state or other form of organisation, political, economic or social therefore has a constitution or is in need of a constitution is taken as self-evident. However as mentioned above, not all nation-states, and in particular the common law jurisdictions, have constitutions. Gordon (1999, Introduction, pp. 1–19), on whose work the following discussion is partly based, views this paramount authority within an autonomous polity as the state, which is the apparatus of government and the individuals that use that apparatus within that governmental structure to exercise power, but distinct and separate from the general population that constitutes the majority of the nation or social polity. The state is then seen as either the source of all evils, or a fundamental social organism with a moral purpose transcending that of any individual. In the latter case, the needs of the people for collective good(s) are provided for through the state, making it a useful instrument for decisions on collective matters. The state also organises society through coercion and is therefore that "institution in society that possesses monopolistic authority to employ legitimate force" according to Weber. Although the source of law is the state, public policy determining law is often developed through the participation of many non-state actors including the wider population, allowing the use of institutional mechanisms and the protection of rights and interests of minorities.

Even a legitimate government, one that perhaps derives power from the people, as in the American Constitution, must be constrained. Laws made by law making authorities should apply to everyone. Constitutionalism therefore implies that the power of the state is constrained. In a social polity, the community exercises aggregate power in dealing with its natural environment, for attack or for defence. However, constitutionalism concerns political power distributed within a community and involves institutions such as the state, church, businesses, research institutes, public interest associations or other corporate body.

A constitution pre-supposes that the group drafting the constitutions represents the values of the collectivity, but not of any one individual, begging the question of where and how this representation comes about, and how representative it is, again a question of the legitimacy, origin and source of the authority involved. There is presumed a contract between the collectivity and the framers of the constitution, this collectivity consisting of a populace that agrees and accepts the values within the constitution, or which can be coerced into such acceptance, or which may require the constitution to be enforced upon them. The drafting and existence of a constitution does NOT pre-suppose democracy in its modern representative form; although the framers of the constitution may believe themselves to be representative, they have generally not been elected or chosen except amongst and by themselves. Further, having a constitution requires a nation-state (Rabkin 2005), with territories and a population willing to accept the constitution. This territory then needs to be defended against those within and without who disagree with its provisions, and also requires internal mechanisms of enforcement to ensure that the behaviours, procedures and relationships enshrined within it are being followed. Apart from such internal institutions, any such polity requires the negotiation of relationships with bodies external to it, those who differ in their values from those of the central constitution. The nation-state also defines who belongs, through subject-hood (subjection), citizenship or similar, through the constitution, thus excluding all those who do not share in those values. It should also be noted here without further comment that the idea of a set of values shared amongst a particular group is similar to that of the ideologies and religions that suppose a chosen people, and is by its very nature inimical to the concept of equality.

There are limitations on what can be part of a constitutional document: is the constitution a description of what actually exists in human society, a descriptive approach, or is it a set of principles from which further principles for implementation may be derived, a prescriptive or normative approach. A descriptive constitution leads to a static and possibly stagnant reading of the content of the document. Amendments responding to change in society and in public opinion would be difficult to incorporate or include and implement in the original constitution and its symbolic, written representation. In contrast to this, prescription or normativity allows principles to be directly applied to a given situation, and allows the derivation and statement of laws applicable to novel situations, not conceived of in the original constitution. In other words, the constitution contains a set of guiding principles for the use of legislators and for the judicial system, both of which institutions may be defined and allowed for in the constitution. However, care should be taken in extending this too far, as a prescriptive constitution is essentially a value judgement about what is considered acceptable and therefore ought to continue to be considered good, right and proper; herein lies the appeal to privileged positions for religious, political and other ideologies and positions. A constitution responsive to changing situations may be better based on normative principles, rather than a snapshot descriptive of currency or contemporary conditions, or on prescriptive principles reflecting values and beliefs not representative of populations. Note here the difficulty of balancing majority with minority opinions.

In general, the modern constitution is presented in a written form in contrast to oral tradition, which symbolic document is then disseminated and placed in the archive of record. The constitution has to be agreed and ratified before implementation by the sovereign legitimating authority specified within the constitution itself, which may be a monarch, the "people" or a representative body such as an elected or appointed parliament amongst the many variants.

The document has a generic structure in most modern constitutions: it starts with a title, followed by a preamble. Although not always the case, the preamble of a constitution is where a generalised and implicit statement of the values and beliefs of the polity are to be found. The preamble is more generally an explanation of the context of the document containing the constitution. This is also often where a statement specifying the legitimating authority is included, as well as the geographical territory, political hegemony and populations over which jurisdiction is claimed.

The remainder of the document is then, if necessary, usually divided into numbered sections with headings descriptive of the primary content. Sections are divided into paragraphs, which articulate, hence also known as articles, a specific concept, principle or procedure, further divided into sub-sections elaborative of the concept, principle or procedure under discussion. Sub-sections, articles, paragraphs and sub-paragraphs do not usually have headings or sub-headings, but are often led and referred to by a hierarchical numbering system, cascading from the section number. These sections can have internal references to other sections of the document, as well as to external sources, whereas immediately necessary material is often placed in appendices or provided as ancillary documentation. Amendments made by due process to constitutions are sometimes added to the end of the main and original constitutional document; in other cases the entire constitution is re-written to present a unique and singularly coherent document.

Whereas the structure of a constitution is simple enough, the content is and can be highly variable. Amongst the most common themes included in any constitution are the value and belief system, usually that of the dominant group of the social polity; a statement of specific rights, responsibilities, duties, obligations and prerogatives; and detailed instructions regarding institutions and procedures, which can also articulate the above stated goal of constraining the government. Sometimes implicit, beliefs are convictions or assumptions generally held to be correct or true, despite an absence of or possibly against evidence. Religion is one such, but all ideologies and unsupported assumptions are beliefs. Values are generally the worth that one gives to things deemed important or essential, usually abstract ideas with applications to concrete situations, such as conservation of the environment, or freedom of speech. These are often internalized by an individual or society and may differ over time, reflecting the need to amend a constitution. Institutions may be the cabinet, an authority leading a government such as a prime minister or a monarch, the various departments of state, the police and armed forces, the church, electoral systems and procedures and the like. These can be created, destroyed, replaced or changed as required subject to the constraints of the constitution. Rights, responsibilities, duties, obligations and prerogatives are a set of ideas that vary from culture to culture, and even over time, and involve a set of reciprocal relations that can confuse and conflate the best discussions. Possible objectives include the protection of individual freedom and the civil rights of minorities, thus serving as a device for peaceful communal association. A good review of these concepts from the point of view of the United States, including the distinction between natural and human rights from a legal and anthropological perspective can be found in Donovan and Anderson (2003, pp. 144–164), to which the reader is referred.

As discussed, legitimation of authority requires a sovereign authority in every stable state. For Aquinas, secular authority derives from the people, and in the Justinian code, the emperor receives his authority by delegation from the people. The "people" are here seen as the proper sovereign authority, a normative assumption. Popular sovereignty remains a "prominent and honoured doctrine" expressed in the "common opinion in democratic states that the best form [of government]… is self-government… achievable if sovereign authority rests at least ultimately, with the people" even though government is at best "the exercise of power by the few over the many" (Gordon 1999, pp. 33–34). However, large populations residing in extensive and varied geographical localities require a form of representative government, as direct democracy by the people is difficult, if not impossible, the legislature replacing self-government as sovereign authority, with the majority, or some sub-set thereof ruling over the whole. Sovereignty can thus reside in many places: the state, the nation, the cultural community, the "law", "reason", the judiciary, the collective general will of Rousseau, Parliament and so on, emphasizing that the "people" is a rather hazy concept: Athens restricted participation by excluding women and slaves, and as in the Roman republic, to select classes of propertied citizens. Reification of the collective will and capacity of the people to make and hold value judgements leads to the modern notion of "by the people, of the people, for the people", the people as supreme legitimate authority, exemplified supposedly in the American Constitution. Popular sovereignty also led to the excesses of the French Terror and of Nazi Germany.

In the Roman Republic, The Twelve Tables extended this by the assertion: *salus populi suprema lex esto* (the welfare of the people must be the supreme law). However, anything can be declared to be law, emphasising the need to constrain the state. Thus, a necessary part of any constitution has to be to specify mechanisms that facilitate this restriction. In establishing a federal government the American system divided legislative powers at the level of nation and state. This federalism distributes power, so there can be no unified seat or locus of' sovereignty. Therefore, according to Dicey, there is no definite locus of legal sovereignty for countries such as the USA; for others it is the legislative body, being Parliament in the UK (Gordon 1999, pp. 45–46).

In most contemporary examples sovereignty is hierarchical, each element subordinate to the one above with a persistent chain of command and an ultimate seat of authority at the top. Where a clear, unambiguous chain of command is necessary, this allows an effective administration and implementation of policy, separate from the actual proposal and determination of policy, and which is most effectively seen at sea. On the other hand, many societies have constructed political systems on the principle of checks and balances or countervailance, the best-known application of which is the separation of executive from legislative powers. In complex pluralistic political systems, power is dispersed across various institutions and in the formal government; independent entities in a network interact with each other, with no supreme authority but with numerous centres of power, the pluralist structure being especially suited to heterogeneous populations. In these pluralist societies the independence of the individual is organised through other non-state and social institutions e.g. religious institutions. Countervailance has an analogue in many disciplines, where a number of independent entities interact so as to maintain a stable system (some in equilibrium) without any one being dominant (Gordon 1999, pp. 17–18). In modern democracies, the state has become identified with the nation and with the people considered as a whole. The state is thus the highest authority, but this does not make it either sovereign or absolute (Gordon 1999, pp. 57–58).

11.1.2 Space Constitutions

In discussion about a series of workshops held in 1985, 1986 and 1987 at which the attendees drafted a *Declaration of First Principles for the Governance of Outer Space Societies*, Robinson (1988, 2004) sees a space civilization forming. In the process, cultural recidivism is rampant, being used to establish the legal foundations and social constructs available off Earth, based on Earth's indigenous historical values, principles and motives, however useless and irrelevant they may be to new societies and civilizations formed in hostile artificial or alien life support environments. The 3-part document has a preamble stating the reasons for the document, a reaffirmation of fundamental human freedoms and the inalienable rights of individuals who live in space, and a restatement of the principle that the governance of and by space societies should reflect the will of the participants. The declaration is designed to evolve and adjust to changing realities (Robinson 2004).

Given the geography and make-up of the participants, and in accordance with the Outer Space Treaties, it is unsurprising that in the first Article of two the authority of the United States Constitution is extended to "all individuals living under US jurisdiction" in outer space followed by statements reaffirming standard constraints on authority protecting personal freedoms, and guaranteeing certain rights, the former including the right to reasonable privacy, freedom from self incrimination, freedom from unreasonable intrusion, search and seizure, and freedom from cruel and unusual punishment, the latter consisting of freedom of speech, religion, association, assembly, contract, as well as the rights of petition, informed consent and private ownership of property. In contrast to Earth-bound constitutions, imperatives of community safety and individual survival within the unique environment of outer space are also recognized and guaranteed, as is travel to, in, and from outer space, and access to media and communications (Robinson 1988).

The second and final Article vests authority in the will of the people of those outer space societies, which societies are allowed to defend themselves, whilst inheriting and assuming all rights and obligations for treaties and international agreements to which the US is signatory. The US also ensures that all petitions originating from the space societies must be heard, including that for an orderly and peaceful transition to self-government when required. Again in accordance with the Outer Space Treaties, science and technology is privileged for the benefit of mankind, with the outer space society further required to "protect the environment and resources of Earth, space, and beyond" (Robinson 1988).

In privileging the rule of law, Natural Law (here remaining non-specific) is the basis for all positive law in this Declaration; this gives humans rights and responsibilities in any environment by virtue of being born. As such, all natural rights and worth "must be premised on those who live, work, formulate and implement the social order of their outer space society, which is developed and exists in an alien environment different to that of Earth" (Robinson 2006). Although, the need for survival in space constrains and balances personal freedoms, the need for survival of a micro-society in a hostile environment can also necessitate the temporary suspension of these rights (Robinson 1988, 2006). Newer essential rights specific to the new environment are those of travel to, in, and from space and access to media and communications, essential for collective and individual safety, and for informed consent when necessary in the presence of the controlled dissemination of information and of continuous surveillance. Ownership is ingrained in natural law and in capitalist economics, therefore property rights have also been included, despite a possible conflict with the Outer Space Treaties (Robinson 1988).

11.2 Artificial Habitats and Facilities: Space Stations

Libra (World Research Inc. 1978) is a film that acts as a manifesto and immigrant brochure for not just an artificial, privately-funded, self-governing space colony, but for its unashamedly libertarian ideals. The Earth of 2003 is described in its initial voiceover:

> Let's face it. Your world is falling apart. Politicians engaging nations in wars against the will of the people. Increasing worldwide poverty and starvation. Inflation, high unemployment, staggering crime rates. Skyrocketing costs of nationalized health care. Overpopulation. Inability to meet your energy needs. Bankrupt cities, bankrupt states, bankrupt nations and morally bankrupt people.

In space, on Libra, the artificial habitat established in the fictional 1978 for only $50 billion in private funds and in a time of looser government regulations, there is hope for freedom and private enterprise, away from micro-management by the world government on Earth. The film describes the fact-finding missions of an investment banker wanting to invest in the power and space manufacturing industries, and of a world government senator wanting to rein in renegade capitalists, these last threatening the world government's monopoly and regulatory stranglehold on the world's energy supply. A film within the film explains that residents are free to "work, raise families and enjoy living", away from the dystopia of Earth, inside a rotating central sphere with a population of 10,000, where the residents exercise their "freesponsibility" to make their "freecisions". Modern reviewers have suggested that the ethos of the film is that of Ayn Rand in her novel *Atlas Shrugged* (Novak 2013). The film is readily available at the film archive website *AV Geeks*. However, Libra is not the only artificial arcology under consideration.

11.3 Interstellar Arcologies: Republic of Minerva on the Starship Aurora

The Republic of Minerva on the Starship Aurora is an interesting example of the outward-looking tendencies of the human race. Apparently an extension of the idea of settling a republic on an artificial island among the Minerva Reefs in the Pacific, it extends the concept of international waters to the interstellar medium. The physical structure of the Starship seems to be composed of a series of islands within the structure of the spaceship. These islands have come together to form nations-states, one of which is called the Republic of Minerva.

The Starship Aurora as Pon explains in the accompanying documentation to the Minervan Constitution (*Starship Aurora* 2001-2) is also a product of the mind of Ayn Rand, written after the Apollo Moon landings (Rand 1969). At first a purely philosophical construct, this abstract idea of a Starship was later extended to imply a physical environment along the lines of space habitats as envisioned by O'Neill and co-workers in the 1970s for a population of 10,000. There is an extensive documentation and discussion dating from the 1970s to the present day archived and maintained online by the founder, Monart Pon, of the Starship Aurora project, which also includes a forum for current discussions. Pon's ideas (available at *Starship Aurora* 2001-2), in particular the technological aspects of this project, appear to have first been articulated in his Master's thesis, and in discussion as recently as 2001 with a consultant, and for whom a Starship is "an integrated [steerable] mobile environment that provides man with nourishment for growth, shelter against decay, and locomotion to explore [other regions of space outside Earth's orbit]" [*sic*] which is a "self-generating, self-sustaining, navigable, high-occupancy environmental system" optimized without disadvantages for human life. This needs an industrial and social infrastructure-complex allowing mankind to attain the stars. Known as "astronautics", this is a "private affair … for private gains at private capital and risks. Any public involvement … does not have any special rights superseding individual rights." Legislation against the free exercise of astronautics is disallowed, and co-operation and organization is encouraged provided government is not used to gain advantage. In sum, the entire thrust of the website, and the Starship Aurora project, which is described as an "Objectivist Society", is to encourage and develop the technologies and human development that allows off-Earth travel.

Pon states that a reader of his original Project Starship pamphlet sent him a bound photocopy of a 20-page document containing a proposed constitution for the Republic of Minerva. The dating is uncertain, but from partial archival records and comments dated 2002 posted on the site by Pon, it would appear that the likely date of the composition of this document is around 1977, when the original pamphlet was published.

The *Constitution of the Republic of Minerva* consists of a preface and a glossary before the preamble, which is followed by eight Articles, each with sections, subsections and explanatory paragraphs. The preface describes the territorial jurisdiction of the constitution, which covers many scattered land areas and islands, each self-governing but with strong ties to central government. The glossary defines 15 terms, which presumably would otherwise lead to misunderstanding. Of particular interest are the definitions for capitalism, for a constitution, and for a utility system, these being central to the ideology of the ship, of its governance, and for the equal access to resources required in this environment. The first, capitalism, is a system in which all personal rights are protected, the second defines a constitution as a set of fundamental principles and laws guaranteeing capitalism, both of which closely ties this Republic to the tenets of Positive Objectivism and the philosophy of Ayn Rand. Utilities include water, energy sources, and communication systems, but not food, air or other supposed essentials, despite the closed nature of the interstellar arcology.

The preamble states the central precept of the Constitution, which is that government exists only to "protect its citizens from force and fraud", a clear statement of extensive liberty and of economic intent. The first Article describes a standard separation of powers in the triply structured government, with a bicameral legislative, a tripartite judiciary and an executive consisting of an elected panel with a chair at the head. The detailed composition of these representative bodies is also described, along with procedures for selection and limits on remuneration. Article II specifies the times, procedures, the required quorum and the rules regulating assemblies, as well as those for impeachment of governing officials in any of the bodies specified in Article I. Article III limits the duties of the bodies in Article I, those of the Council being budgets and the distribution of funds, but also to negotiate international agreements. The Executive Panel is limited in its powers to act for purposes of security, and in case of emergency and related legislation. All legislation, emergency or otherwise is annulled after a specified period, ostensibly to curb a proliferation of unnecessary and interfering laws. The function, organization, financing, use, discipline and administration of a military force are covered in Article IV, members of the armed forces being made to swear an oath of fealty.

This oath, as does the need for strong ties to central government, appears to contradict both the free-market approach and the freedom and liberties generally demanded by an Objectivist approach. Article VI covers fiscal matters. Whereas suitably backed currencies can be issued by private citizens, government is not allowed this function, whilst central funds are sourced through taxes ("premiums") and court funds, or through "the registration of ships", a phrase which is unclear in this context. Article VII limits the application of the legal code by the judicial system of court attorneys and judges to "all persons, regardless of race, religion, origin, wealth, or influence" on the basis of equal rights, any acts involving force or fraud being considered crimes. Articles VIII and V are especially pertinent to this analysis: these limit the activities of government in relation to the Republic and its citizens, and with respect to foreign relations respectively, Art. V restricting the government to diplomatic relations involving trade, the acquisition of land from other polities, or for purposes of extradition or immigration. The first half of the former constrains government to administrative functions according to specified procedures, covering patents and copyrights, and the exploitation and distribution of water resources. The second part of Article VIII details personal responsibilities with respect to minors in given situations. It goes on to state principles of ownership, (equal) rights to water and to mineral resources, and a complete prohibition on the contamination and pollution of air and water where this might constitute a danger to persons, plants, or animals found elsewhere.

11.4 Planetary Settlements: The Tharsis Montes Commonwealth

Written by Paul Burgess, a pastor in the United States, in 1987, this constitution is a response to a confirmand who had been assigned the task of composing and writing a constitution for an off-Earth colony as part of a school assignment. Burgess has posted his original constitution without comment or changes on his website (Burgess 1987, posted online 2001). Burgess appears to have an interest in science fiction at least, and from the content of the constitution, a fair knowledge of both constitutional government *à la* USA, and an interest in space colonization.

Burgess has provided a comprehensive contextual background, partly in the Constitution itself and partly in an appendix. The *Constitution of the Tharsis Montes Commonwealth* consists of a title, 6 sections, each with several paragraphs, and an appendix on demographics at the end of the document. The six sections start with the equivalent of a preamble, then goes on to describe the legislative body, the executive body and the judicial system in the next three, with basic civil rights as protected by the constitution, and procedures for amendments to the Constitution discussed in the final two sections respectively. At the end, but before the appendix are appended the signatures and affiliations of the signatories, along with the date of signing.

The first section specifies the name of the sovereign unit in the first paragraph, and then goes on in the second paragraph to declare the legitimating authority, a fictional UN treaty under which it acts as trustee on behalf of the fictional Luna, Tharsis and Vesta Development Corporation operating on the Moon, Mars and the asteroid Vesta. The third and final paragraph delimits the extent of the territories over which jurisdiction and ownership is claimed, these being a geographical area specified by longitude and latitude on the planet, any settlements away from the central metropolitan territories, orbital habitats, platforms and space transportation vessels. Jurisdiction is also claimed over any material and manufactured goods originating from the Commonwealth or its citizens whilst in transit to its destination. Of interest in the final sub-paragraph is the provision for the inclusion of any territories, settlements and equipment obtained "lawfully" in the future by the polity.

Section 2 describes the Citizens' Assembly as the locus of sovereignty, being granted the power to legislate by virtue of paragraph 1, and in the subsequent paragraphs 2 and 3, the eligibility for which is limited to citizens, these being those registered at the time of chartering of the Commonwealth, those born in the territories registered once having attained a majority of 14 years of age, and those naturalized after a total of 8 years of residency and public service for 6 and 2 years respectively. Citizenship may be bought, and also be awarded for public service. Although citizenship may be denied or withheld, whether or not the applicant is native, the reasons must be made public. The non-native candidate upon refusal must also be provided with the next available passage to Earth equivalent to the value of assets accrued during the candidate's qualifying residency period, with some exceptions. The charter mentioned does not appear to imply independence from its originating body, but possibly refers to its establishment.

Paragraphs 5 and 6 of the second section describes the procedures and quotas required for introducing legislation, through sponsorship by a specified proportion of the populace, through initiative of the Commonwealth Council or through a duly constituted Public Forum, the procedures for which last are specified in paragraph 10. Proposed legislation requires a simple majority to be passed and enacted, the schedule for implementing legislation being also specified in this constitution, with voting freely enabled through the computer networks maintained by the central government.

Some important exceptions are specified regarding legislation in paragraph 8, these being an absolute ban on retrospective legislation, on attainder where all civil rights are forfeit and removed from the person in question whatever the crime, and on the corruption of blood, where a person would lose all rights to inheritance, including land, and to retain or transfer such properties to another person.

Executive powers as described in the third section are vested in the Commonwealth Council which consists of a majority of citizens who are independent or represent public interest bodies, in addition to 3 citizens appointed by the founding Corporation, and one representative respectively for each of the professional and trade guilds present in the Commonwealth, with elections held every 2 Martian years and candidates being at least 16 terrestrial years of age with a subsequent 2 years of recognised public service. The Council, whose Chair is chosen from within its membership, is responsible for enforcement of legislation passed by the Citizens' Assembly and those decisions passed down by the judiciary, can react to emergencies for a limited period without formal approval, and appoints judges. Finally, the legislative body may impeach and punish any Council member, although loss of citizenship and deportation from its territories is the maximum allowable penalty.

Judges for the judicial system are nominated and confirmed by the executive and legislative bodies respectively, the procedures in this section being dependent upon and yet independent of the preceding two, with a single judge being necessary for civil cases below a threshold value, and for criminal misdemeanour; felony and capital cases require adjudication by three, and all capital convictions are subject to review by an independent panel without right of further appeal. However, any conviction for felony leads to removal of citizenship and associated rights and privileges, with the discretionary possibility of passage to Earth, the confiscation of all goods and assets and probable confinement until departure. Judges may be impeached, are required to reach their decisions based on evidence and testimony "according to strict constructionist principles", and are not permitted any judicial review of legislation. Judicial jurisdiction does not apply for the duration of emergencies, nor extend over authorized officials for actions in the performance of their duties "in free fall, in orbit, or in deep space, or while under pressure suit conditions."

Section 5 sets out the civil rights accorded the citizenry. Paragraphs 1 to 3 make explicit that religion may not affect any aspect of a citizen's rights, privileges or other capacities., nor may freedom of speech and self-expression be limited in any way except by reference to "a clear and immediate threat to public safety", and that the legal right of residents to peaceable assembly is subject only to the "physical confines of available space and the capacity of support systems".

Paragraph 5 of this section is especially interesting in that the Commonwealth holds "all real estate, air, water, and food rights" in trust, further "entrusted" for "use", but not "ownership" by citizens, residents and visitors, subject to the payment of taxes as determined by the Assembly. Further, a citizen retains trusteeship over resources administered by him except in cases of felonious behaviour or of tax arrears. In contrast to this, according to paragraphs 6, 7 and 9, outside of fraud and illegality, all citizens have the right to hold capital, goods and enter into binding contracts, and the people to be inviolate in their persons, houses, data and effects, and against unreasonable searches and seizures, with the right to a speedy and public trial where relevant.

Paragraph 8 ensures that events in "free fall, in orbit, or in deep space, or while under pressure suit conditions" or cases involving a "clear and immediate danger to the life support capability" of settlements remove the right to withhold testimony regarding one's own involvement in a felony, presumably to avoid incriminating one's self, whereas a refusal to provide such testimony allows travel to Earth "without loss of citizenship, capital or goods", thus avoiding such self-incrimination.

Finally, paragraphs 10 and 11 in this section state that rights not specified in the Constitution are still retained by the people without prejudice; similarly powers not specified are reserved to citizens. However, where a "clear and present threat to the life support systems of the settlements" is evident, no part of the Constitution may be used to avoid any necessary action to deal with the threat. As such, presumably all rights are suspended for the duration of the emergency.

Amendments to the Constitution are allowed for in the sixth and final section and are subject to similar restrictions and procedures as those specified for the Citizens' Assembly in proposing and enacting legislation.

In the appendix on demographics, Burges provides a fictional account of the distribution of humans in the solar system in 2043, as well a fictional history of the Tharsis Montes Commonwealth, which is only one of several colonial polities on Mars. It consists of the main city, several smaller satellite settlements, nomadic and other encampments and one orbital platform. Of particular relevance here is the role of the fictional Aristarchus-Center-based Luna, Tharsis, and Vesta Industrial Development Corporation which sponsored the Tharsis Montes Commonwealth sovereign charter under a fictional United Nations treaty, providing significant financial backing and exerting substantial political influence in the fledgling Commonwealth.

11.5 Earth: *National Space and Aeronautics Policy Act* of 1981 (USA)

In 1981, Gingrich and his colleagues sponsored a private bill, HR 4286, presented to the 1st session of the 97th Congress in the House of Representatives to "establish a national space and aeronautics policy, and for other purposes". The bill was not passed but was referred to a House committee and subsequently died, never becoming legislation. Also known as the Northwest Ordinance for Space, it is Title IV of this bill that is of relevance to this essay, focussing as it did on the "Government of Space Territories." The first section of this Title provides for constitutional protections in requiring that all persons residing in any US space community "shall be entitled to the protection of the Constitution of the United States." In its second section, Title IV allows any US space colony to hold a convention that may establish a constitution and to decide its form of self-government as long as that colony has a minimum population of 20,000. Complementing this is the official transition to the colony being admitted as a full state "on an equal footing with the original states" into the US federal system provided that the population of the colony has reached a threshold not less than the population of the least populous US state at the time of the petition for entry. The remainder of the bill is concerned with US involvement in space exploration and exploitation at the least, and is not of immediate relevance to this essay on constitutional governance beyond Earth.

11.6 Discussion

11.6.1 Comparison of Constitutions

The selective nature of the data limits the comparisons that are possible, given that all examples examined in this essay are derived from and heavily influenced by the American Constitution. This also reflects the history of space exploration until the late eighties, generally being restricted to the two nations of the USA and the then Union of Soviet Socialist Republics. However, it is apparent from examination of the statements in the *Libra* manifesto, the Minervan and Tharsis constitutions, and the proposed US legislation that several essential elements of each are the rule of law, pursuit of wealth, and of liberty and freedom. Wealth is evidenced by the power and manufacturing industries in space, by the antecedent corporation that set up the initial settlement at Tharsis, and the protection from fraud for the Minervans, as well as the proposal to exploit off-Earth resources in the 1981 Act. All four presuppose, or explicitly require constraints on authority, and re-affirm the rights and liberties due an individual. And the rule of law is both consequence and pre-requisite for the acceptance of a constitution. The final element that enters into this is a consequent deriving from the American Constitution, that of the pursuit of happiness. Taken together, this corresponds closely to the four values stated in classical liberalism in the Century of Enlightenment, and are similar, if not identical to the Vedic requirements for a successful life: these being law to structure life, the acquisition of wealth, the fulfilment of desires leading to happiness, and the freedom from constraints that allow the pursuit of the other three goals. All examples studied privilege differing forms of free-market capitalism.

The proper seats of authority, and therefore sovereignty able to exercise jurisdiction over a population and its territories are likely to vary in each given situation considered, namely, an artificial habitat, an interstellar arcology, a planetary settlement and the Earth nation of the USA. *Libra* is a self-avowedly independent and sovereign polity, under no jurisdiction declared in the film. The society is Libertarian in its ideological principles, and presumably it follows that the locus of sovereignty therefore lies with its inhabitants, these being encouraged to be responsible for their own decisions. A different form of Libertarianism is the Objectivist ideology of the Republic of Minerva. The word republic implies that it is a thing of the people, and by extension for the people. There is no linguistic or historical derivation possible for extending the meaning to rule by the people, except to point out that Greek and Roman republics were generally oligarchies and aristocracies. This is generally in line with the strong central government links required to co-ordinate the scattered sub-polities, but inconsistent with the existence of a well funded and constitutionally essential military force required to keep the citizenry in line, and in contravention of the requirement of minimal interference. The "Constitution [is to] enable the effective governing … … ties to the centr[e] are … strong, with […minimal…] interference". This coercion suggests an authoritarian approach is in place, which counteracts the claims of freedoms and rights for the population, but which presumably supports an economic autarchy. The forces of the free market displace the authority vested nominally in a government by the people. It is therefore likely that the essential seat of sovereignty is now the invisible hand referred to by Adam Smith.

Further, given that the Republic of Minerva is situated on an artificial habitat, like the *Libra* space station, it, and sections of it are subject to ownership with associated property rights, and jurisdiction therefore exercised as an independent and sovereign polity without reference to its originating public or nation-state. In contrast, the Tharsis polity is constituted and named a Commonwealth, etymologically suggesting the wellbeing of the common people. The Commonwealth is situated on a celestial body, which although not technically or legally defined in the Outer Space Treaties, is assumed to mean planets and moons, and presumably dwarf planets. It would therefore have been originally subject to the jurisdiction of the originating body, delegated to the founding corporation, which then hands down its jurisdictional powers to the settlement. "Tharsis is a sovereign governmental unit … serving as trustee … for the Luna, Tharsis, and Vesta Industrial Development Corporation". At this time, however, the settlement cannot be considered to either independent or sovereign in its people, but only as a delegated authority. Finally, members of the Republican Party sponsored the 1981 Act under US jurisdiction, underlining its free market and mostly Libertarian principles.

Both Libertarianism and Positive Objectivism look inward to and exalt the egocentric and rational individual above the collective, in contrast to the first modern constitution that invokes a government of the people and by the people; however, this, the Constitution of the Commonwealth of Connecticut in the 17th and 18th centuries (*History of the Connecticut Constitution* 2014), was designed with regard to the welfare of the people as a whole, and not the individuals from which that collective is composed, a perspective of direct relevance to the hostile space environment within which the bio-fragile humans must survive. Nor do these ideologies recognise that the infrastructure necessary is built and maintained by collective action and government.

In addition, Minerva privileges the ideology of Ayn Rand in its Objectivist form, which has therefore displaced the implicit position held by religion in Western nation-states where references to "God" derive from a legacy of both regulation and legitimization by the Christian Church of the state and its rulers. Minerva further privileges the economic ideology of capitalism in its generally recognised form, with an explicit reference and definition provided in the glossary, this being part of the written document.

11.7 The Outer Space Treaties and Sovereign Nations

Traditional notions of sovereignty would have been central to the appropriation of property by nation-states in the Outer Space Treaties of the 60s and 70s. Government requires a state with a stable political system that is hierarchical, sovereign and unconstrained, with the state conflated here as both an autonomous jurisdiction and as its own supreme legislative authority. Government by constitution is necessarily constrained, but the required state remains hierarchical and sovereign. Sovereignty as traditionally understood is absolute, indivisible and permanent (Gordon 1999, p. 21), can be delegated but not shared; the distributed sovereignty of federal systems, or of subsidiarity and paramountcy practised in the Indian empires of the British, Mughal and Hindu kings or in the European Union would not be considered sovereign, whereas a monarch with a parliament or a ship's captain would. Sovereignty must not contradict any previous law, precedent or contract, these remaining in perpetuity. Its indivisibility is why church and state remained tightly locked together in Europe, whereas written texts show that the religious was separated from the secular in the Indian sub-continent before 800 BCE. Sovereignty is not therefore always indivisible. This has significant consequences for states in outer space when related to the Outer Space Treaties.

The Outer Space Treaties do not allow the sovereign national appropriation of property on celestial bodies, or of empty space, the latter in particular being treated as part of the common heritage of mankind. If such ownership were allowed, then the only route to a sovereign statehood would be an outright declaration of independence from the sponsoring body, as any form of federalism would create shared or ambiguous ownership, unacceptable since constitutional governance requires a nation-state that is sovereign. Closer examination of the OSTs show there are no restrictions on sovereign states existing in space as long as these do not derive from Earth based jurisdictions, thus bypassing the national sovereign appropriation clause, assuming that the clause probably applies only to claims arising from sovereign jurisdictions based on Earth, and that the obligations of the international treaties are not inherited or assumed by the new state. Consequently, any entity wishing to be seen as sovereign must declare itself autonomous and independent of its originating jurisdiction as a necessary step for sovereignty and constitutional governance, unless a federal system is instigated. However, a federal system would probably also be disallowed under the terms of the OSTs, given that there is sharing and delegation of an Earth based sovereignty.

Although no ownership of landed property is possible, there is no restriction on the use and exploitation of resources. Jurisdiction over materiel and personnel in space is the responsibility of the originating nation-state. Jurisdiction by a sovereign state through an extended form of usufruct is a possibility under the Space treaty system. This jurisdiction can then be delegated to or shared with the settlement through either a federal system of some sort, through a form of subsidiarity, paramountcy or federalism, and eventually through, although this is open to debate, independence. Ownership without property rights under conditions of usufruct is similar to that of nomadic pastoralism and the use of common land, where there is no single or plural corporate or individual body that has ownership or other unique rights to the property over the territory, buildings, inhabitants or its produce and resources. There is therefore a collective use of the territory and its resources, but no collective ownership, by those requiring it, but which does not exclude use by others. This brings in the idea of a commons, or common heritage, but not of common ownership, as there remains no property to own, differing slightly from one standard legal definition of usufruct whereby a person is allowed use of property owned by another. The founding state continues to have jurisdiction over the colony, which can claim neither ownership, nor sovereignty unless independence is declared or given. The space treaties also make no provision for the exercise of concurrent jurisdictions by one or more states in control of a settlement accommodating non-nationals. In the International Space Station currently, jurisdiction in each module remains with the maker, launcher &c. of each module.

An alternative is that at some future date, the Outer Space Treaty system is no longer considered valid, the principle of *rebus sic stantibus* being applicable, being "a tacit condition said to attach to all treaties that they shall cease to be obligatory as soon as the state of facts and conditions upon which they have been founded have substantially changed" [Latin, at this point of affairs; in these circumstances.] (Robinson 2006), which would completely change the discussion.

11.7.1 The Road to Statehood

Recognising that it is not possible to resolve the question of landed property rights under the current system of the Outer Space Treaties, there is no bar to occupancy and therefore use with respect to the principle of usufruct although there is no owner in this instance. This raises the question of whether the settlement that occupies the land without ownership, not having rights to it as property, can be declared independent, and ultimately sovereign, and how the question of sovereignty is to be resolved. If the settlement is such that it allows an extension of its seat of authority to an external source, such as a founding corporation, an originating nation-state or the like, and that it recognizes this as one of the many sources of authority valid within its jurisdiction, it is possible for the settlement to be declared independent of the originating jurisdiction by that self-same external jurisdiction, and therefore sovereign. A state being necessary for constitutional governance, as we seen above, such shared jurisdictional sovereignty is seen as an equivalent to a system of internal checks and balances, but with one or more components acting from outside the polity. This is reminiscent of subsidiary, paramount or federal systems as discussed elsewhere in this essay.

Then, to form a state requires a mechanism for such transition. In the early days of Connecticut State (*History of the Connecticut Constitution* 2014), the power of self-government and statehood was awarded by the monarch and by royal charter, allowing governance by constitution, a series of fundamental orders establishing the framework of government. In the absence of a monarch or of royal charter, the power to confer statehood may be contained in the antecedent legislative body that exercises jurisdiction. The 1981 Act in the USA discussed above contains one such mechanism, allowing a transition to inclusion within the federal system of the United States provided certain conditions are met. Note here a possible conflation:

a state within the US system is not necessarily a nation-state, which discussion is best left to another day. There exist other mechanisms for establishing a separate social polity of course: a referendum, a revolution, some other method specified from the start, a constitutional transfer or devolution of power. The habitats and settlements presented above, apart from Tharsis, do not consider any mechanism for autonomy. Both artificial habitats, the space station and the interstellar arcology, are presumed independent and sovereign from the start, and any political entity within a starship is also deemed sovereign, albeit only in relation to other equivalent polities on board. The planetary settlement pre-supposes some unspecified method of transfer of power under the aegis of some as yet non-existent global government, subject further to the provisions of international treaty. Such a transition to autonomy may be mediated by a "document leading to an ultimate constitution for space land societies and consequent unique civilizations". This Migratory Manifesto identifies, delineates and defines the rights, duties and expectations of both those on Earth and "members of long-duration space societies during the early phases of such migration leading to a space civilization", hoping to avoid dissension and conflict resulting from cultural and political recidivism (Robinson 2006).

An alternative method of such transition is to rebel, or revolt against the jurisdictional authority. Constitutional sovereignty retains at least the power to make law, or to coerce through the use of or without laws (Gordon 1999, p. 28). The posited Hobbesian contract between people and sovereign does not include the right to disobey or to rebel against a tyrannical authority, instability being caused by satanic votaries, or in the modern Western capitalist representative democracies world by dissidents, and external influences, Islamic, Soviet or otherwise. For Locke, du Plessis and Maritain, egocentric and rational people form a civil society, which then delegates power and subjects itself to a representative sovereign authority that promotes and protects the "people's" self-interest. However, being as it is irrational to accept an absolute authority over one's person, property and actions, society has the right to rebel against tyranny. Revolution does not have to be just against tyranny, but can be based on any acceptable disagreement, although these generally involve the imposition of taxes, the removal of civil liberties, or economic necessity. In modern polities, citizens retain their individual judgements and do not render automatic obedience to laws passed by the legislature (Gordon 1999, pp. 50, 55). Compare this with the attitude of French Protestants, as well the traditional Hindu populations of the Indian sub-continent, rebellion against tyranny being assumed as a right. The Hindus mentioned above have, for example, the right to rebel or migrate in times of famine and tyranny, among other situations.

11.8 Foreign Relations

Of necessity, whatever the circumstances, sovereign statehood may be obtained by a declaration of independence. The existence of a separate polity in space, amongst many both on- and off-Earth would need to regulate relations between these separate entities. Minerva considers foreign relations to be restricted to trade, extradition and immigration. Being one polity amongst many, there is an emphasis and inclusion within the declarative principles of the constitution of the need for foreign policy, and principles, regulations and procedures to govern these external relationships. In contrast, the Commonwealth of Tharsis has no provision for such relations, except implicitly with its parent, the commercial enterprise known as the Luna, Tharsis, and Vesta Industrial Development Corporation complemented by a somewhat more explicit acceptance of relations with Earth and its various nation-states or its representatives, for deportation in several circumstances. In general and in consequence of their foreign relations, habitats and societies should have free access to interstitial space and celestial bodies and be recognised as independent in their citizenry and cultural and political sovereignty, and not subject to claims by Earth-based nation-states or regional jurisdictions (Robinson 2006).

11.8.1 Membership of a Polity

Note that in modern times, a constitution also defines the qualities required to be a citizen, and thereby restricts who can be or become a citizen or its equivalent, with consequences for access to full participation, resources and facilities within a polity. Tharsis defines conditions for required for qualifying as a citizen, whereas Minerva does not. Although essential services are extended to all residents within the polity of Tharsis, and restricted in Minerva, the question of who belongs is important as rights, duties, responsibilities, obligations and prerogatives are limited in general by constitution to those qualifying for citizenship. In the United Kingdom, for example, all residents are subjects of the monarch, a privilege that can be awarded or removed at the monarch's whim. After naturalization, a non-native is usually awarded this privilege, through the agency delegated by the monarch or by Parliament. It is only as a result of the Treaty of Maastricht that a UK subject is considered to a "citizen" of the European Union, but cannot be seen as a "citizen" of the UK. However, the United Kingdom does not have a written constitution, its system being based on common-law and precedent. Neither does the European Union, the series of Treaties signed and ratified by the representative bodies in each member state being considered an equivalent body of constitutional law, along with directives and judge-led decisions originating in its institutions. Note that none of the constitutions examined consider a single language for its citizenry; on the other hand, neither is the necessity of a single language for communicating issues of safety supposed necessary. This implies that the membership of any of the polities above is monolingual, regardless of ethnicity, culture or otherwise.

11.9 Conclusions

Although there is an overall focus on protecting the people from the state, the discussion above also emphasises the necessity of autonomy; thus drawing attention to the two complementary definitions of sovereignty, sovereignty referring not only to autonomous jurisdiction within the state's own territories, but also the need within each state for an authority that is politically and legally supreme.

The several documents above show that they fall within the accepted framework of the study of constitutions. With the American Constitution as an exemplar, each approach above contains the standard element of defining the seat of authority and therefore the locus of sovereignty in the polity. They also contain the now standard elements of some privileged ideology representing the stated values and beliefs whether implicit or explicit, followed by a governmental structure that consists of checks and balances, in accordance with the principle of countervailance, thus ensuring that the authority does not overstep its bounds. These bounds are further constrained by specifying the freedoms granted to individual person within the polity, and of the rights and responsibilities given to and expected of these same individual persons, comprising what is now a standard list of such liberties expected in Western democracies. However, as mentioned, they reflect primarily American values, and do not take into account the precedent, and now competing and complementary value and belief structures found in present day Europe, the major Asian countries of India and China, nor of the cultures of the Middle East, Africa, South America, or Australasia and the Pacific Rim.

In conclusion, the documents presented above concentrate on democracy and rights in a US style constitutional framework, on the relationship between authority and the inhabitants in the environment of space, countervailance being a guiding principle for the structure of government, applied primarily by separating legislative, executive and judicial institutions. However, it becomes apparent that in relation to Earth-based jurisdictions, the exercise of this governance requires independence and autonomy. Several routes to a sovereign status are suggested both within and outside the purview of existing international law. The documents are further implicit or explicit in their ideological foundations, privileged in the detail contained within the constitution, secular science and technology taking the place of religion, as do Libertarian and Objectivist principles. Further, non-partisan groups recognise space as being hostile to bio-fragile humans, whereas those based on specific ideologies ignore this inimicality.

References

Burgess, P. (1987). Constitution of the Tharsis Montes Commonwealth. Retrieved May 4, 2014, from http://www.paulburgess.org.

Donovan, J. M., & Anderson III, H. E. (2003). *Anthropology & law*. New York: Berghahn Books.

Gingrich, E., Dornan, W., Derwinski, et al. (1981). *To establish a national space and aeronautics policy, and for other purposes* (USA, 97th Congress, 1st session).

Ginsburg, T., Elkins, Z., & Melton, J. (2007). *The lifespan of written constitutions*. Cambridge, UK: Cambridge University Press.

Gordon, S. (1999). *Controlling the state: Constitutionalism from ancient Athens to today*. Cambridge, Mass: Harvard University Press.

History of the Connecticut Constitution. (2014). Retrieved December 21, 2014, from http://en.wikipedia.org/w/index.php?oldid=602764375.

Kelly, T. (2012, September 27). *The Mars constitution*. Retrieved January 19th 2015, from https://prezi.com/fw4sfju2ojyl/the-mars-constitution/.

Mars Live!. (2015). Retrieved January 19th, 2015, from http://vasts.spacegrant.org/modules/10.

Merritt, D. (2010). *Draft Mars constitution*. Retrieved January 19th, 2015, from http://c2.com/cgi/wiki?DraftMarsConstitution.

Metcalfe, K. (2012, October 26). *Constitution of Mars*. Retrieved January 19th, 2015, from https://prezi.com/kdmrvtkqowko/constitution-of-mars/.

Mukerji, S. (1969). *The republican trends in ancient India*. Delhi: Munshiram Manoharlal.

Novak, M. (2013, February 15). *Libra: The 21st century (libertarian) space colony* Washington: smithsonian.com. Retrieved January 11, 2015, from http://www.smithsonianmag.com/history/libra-the-21st-century-libertarian-space-colony-19119736/#dXcMjD7JquEU40WL.99.

Rabkin, J. A. (2005). *Law without nations? why constitutional government requires sovereign states*. Princeton, NJ: Princeton University Press.

Rand, A. (1969). Apollo 11. *The Objectivist, 8*(9), 13.

Robinson, G S. (1988). Re-examining our Constitutional heritage: A declaration of first principles for the governance of outer space societies. *Berkeley Techology LJ J2*(3), 81–97.

Robinson, G. S. (2004). No space colonies: Creating a space civilization and the need for a defining constitution. *Journal Space Law, 30*, 169–179.

Robinson, G. S. (2006). Transcending to a space civilization: The next three steps toward a defining constitution. *Journal Space Law, 32*, 147–175.

Starship Aurora—Constitution of the Republic of Minerva. (2001-2). Retrieved May 4, 2014, from http://starshipaurora.com/minerva.html.

World Research Inc. (1978). *Libra*. AVGeeks. Retrieved January 11, 2015, from https://www.avgeeks.com/wp2/libra-1978/.

Chapter 12
Towards a World Space Agency

Operational Successes of the International Seabed Authority as Models for Commercial-National Partnering Under an International Space Authority

Lewis Pinault

Abstract Adopting key principles from the International Seabed Authority established under the 1982 UN Convention on the Law of the Sea, the author first introduced a proposal for an International Space Authority at the Third United Nations Conference on the Peaceful Uses of Outer Space in Vienna in 1999. Based largely on its intuitive appeal to an international community of young space researchers, this won top-voted support from amongst 160 Space Generation delegates from 60 nations and was subsequently incorporated into the Conference report. Many of those delegates today are leaders and innovators transforming the competitive landscape of space development. Meantime without fanfare the International Seabed Authority itself has become a working institution with practical exploration, mining, science and education successes. In the absence of a similar authority for space, political activism to extend privatisation from space launch services to private ownership of space science data and celestial bodies raises key questions on the nature, meaning, and ethics of human exploitation of resources on- and off-planet. This chapter examines how the principles and practices of the International Seabed Authority might inform an International Space Authority, and ultimately lead to more broadly accepted universal principles embodied in a World Space Agency.

Keywords Asteroid mining · Commons resources · Deep seabed mining · Extraterrestrial liberty · Human governance beyond Earth · International seabed authority · International space authority · Lunar resources · Mars exploration · Moon · Space commercialisation · Space economy · Space exploration · Space law · Space policy · Space science · World space agency

---

L. Pinault (✉)
UCL/Birkbeck Centre for Planetary Sciences, London, UK
e-mail: l.pinault@ucl.ac.uk

C.S. Cockell (ed.), *Human Governance Beyond Earth*,
Space and Society, DOI 10.1007/978-3-319-18063-2_12

12.1 A Changing Space Industry Merits a Revisit of an International Space Authority

In July 1999 as the world celebrated the 30th anniversary of humanity's first setting foot on the Moon, the author presented recommendations for the establishment of an International Space Authority to the Space Generation Forum of the Third Conference on The Peaceful Uses and Exploration of Outer Space.

These recommendations were among the final 10 that were selected for incorporation into the official Conference report, and it afforded the opportunity to work with Conference attendees and faculty of the International Space University to expand and develop the ideas presented, focusing specifically on issues of governance for space settlements on the Moon and Mars (see e.g. UN 1999a, b; UNISPACE III 1999). The International Space Authority was proposed to

(1) Mediate priorities for competing interests in space exploration and development;
(2) Provide for equitable use and access to the resources and knowledge benefits gained from these activities; and
(3) Establish a vehicle for pooling centralised funds and resources for united efforts in these areas.

At the time the author sought to specifically address how the International Space Authority could fulfil regulatory functions mandated by the Moon Treaty (UN 1984), in a way that seemed likely to both greatly broaden accession to the Treaty by current space powers, and strongly validate its role and importance for non-space powers (Pinault 2000a). In essence, this involved a controlled but real role for the private sector in lunar exploration and development, but in the context of equitable regulations and agreements for managing competing priorities.

I began to develop in some detail how the Moon itself is a natural priority for space exploration and development, requiring and benefiting from the structure and activities of an International Space Authority (Pinault 2000b). The Moon, I argued, is of more importance to the immediate future needs and interests of the world as a whole, and provides a far more workable platform for global involvement and benefits, including practicable private sector investment, than any other large-scale efforts in space, be it Mars,[1] asteroid or Earth-orbiting station (there are a wealth of works espousing and detailing the practicalities of a return to the Moon even in the aftermath of President Obama's decision to focus NASA's energies more toward Mars (e.g. Crawford et al. 2012; Crawford 2014, 2015; Crawford and Joy 2014; Bowler 2014; McKay 2013; Anand et al. 2012; Jaumann et al. 2012; Spudis and Lavoie 2011), with Benaroya (2010) hedging bets for Mars too. There was also

[1]As of this writing the author nonetheless is seeking flight status as an Industrial Astronaut with the Shackleton Energy Company for missions to the Moon and Mars, qualified as a semi-finalist in the Mars One programme, and serves as a member of the Lunar Mission One education and science teams—welcoming all opportunities for in situ research! (Lunar Mission One 2015; Mars One 2015; Shackleton Energy Company 2015).

a last surge by lunar development stalwarts in the heydays of the previous administration's stated commitments to return to human exploration of the Moon (e.g. Crawford 2006; Duke et al. 2006; Jolliff et al. 2006; Spudis 2001), all building on important works from the 'darkest days' when the Space Shuttle seemed to waylay Apollo's destiny (e.g. Heiken 1991; Ehricke 1985; Hartmann 1985; Taylor 1985).

As we now near the 50th anniversary of the first walk on the Moon, we can now better anticipate in more concrete terms the next phases of human exploration in space, and they build from a mix of important, sometimes competing, and often unexpected turns:

- Visionary major new investments by the private sector under space agency auspices, most notably SpaceX under contract to NASA (Space Exploration Technologies 2015)
- New national players moving on accelerated paths to space exploration competence, including China and India (Space Flight 101, 2015)
- A Global Exploration Roadmap guiding major space agency decisions (Crawford 2014; ISECG 2013)
- Citizen Science initiatives increasingly powered by social media (Joy et al. 2009)
- Unprecedented public access and outreach through micro-satellite and other distributed technology initiatives (Planet Labs 2015)
- New systems models (Metzger et al. 2013)
- New funding models for private (and possibly public) space initiatives, including Endemol's exploratory interest in reality-television programming for Mars One, wealthy individuals for Inspiration Mars, Dubai Expo2020's partnership with Shackleton Energy Company, and KickStarter for the first phase of Lunar Mission One (Inspiration Mars 2015; Mars One 2015; Shackelton Energy Company (2015); Lunar Mission One 2015)
- Prize competition models (Moon Express 2015; Hall et al. 2013)
- Private direct ventures for asteroid and other space resource development (Deep Space Industries 2015; Lewicki 2013; Planetary Resources 2015; Shackleton Energy Company 2015)

In this new maelstrom of increasingly complex activity, the Moon Treaty's widely perceived over-reach in trying to further embed the common heritage principles of the Outer Space Treaty at the expense of private incentives has effectively pushed it to the boundaries of irrelevance (Crawford 2015; Viikari 2012). Many of these same principles however were woven into the fabric of the 1982 UN Convention on the Law of the Sea (Heim 1990; Hoffstadt 1994), and with the benefit of favourable economics on Earth for extraction of rare earth metals (Chakhmouradian and Wall 2012), the operating principles and practices of its execution arm the International Seabed Authority have been worked and evolved to a level of practical detail that largely surpasses political and ideological debates and succeeds in wins for national governments, private contractors, the science community, and arguably, the seas themselves. Recently there have been some especially thoughtful and well-researched pieces advancing the thinking on getting past the Moon Treaty to practicable and responsible systems of governance

(Delgado-López 2014; Ehrenfreund et al. 2012, 2013; Tronchetti 2009, 2014), and to these insights and explorations I would like to add the practical accomplishments of the International Seabed Authority as a stimulant.

12.2 Regulatory Achievements of the International Seabed Authority Offer More Than a Mere Model for Regulation of Outer Space Activities

It has been nearly twenty years since the International Seabed Authority became fully operational as an autonomous international organization in June 1996, when it took over the premises and facilities in Kingston, Jamaica previously used by the United Nations Kingston Office for the Law of the Sea (International Seabed Authority 2015).

Its main purpose is to award equitable licenses and contracts for exploration and mining to governments and private contractors who abide by the biodiversity and other science-driven constraints of the Law of the Sea Treaty. Its initial focus was on the mining of polymetallic nodules and polymetallic sulphides. While licenses and contracts have been awarded, these minerals remain problematic as economic resources—the first because of the great depth of retrieval (c. 5000 m) generally required, and the second because of the sensitive hydrothermal ecosystems intertwined with them (United Nations 2013; Markussen 1994).

The third category of deep seabed mineral resources under the Authority's remit however is proving more promising. Cobalt-rich ferromanganese crusts principally occur at shallower depths of 400 to about 2000 m in areas of significant volcanic activity. The crusts grow on hard-rock substrates of volcanic origin through the precipitation of metals dissolved in seawater in areas of seamounts, ridges, plateaus, and where prevailing currents prevent deposition of unconsolidated sediments, and they occupy large areas on top of these topographic highs. As many Pacific Rim nations are themselves volcanic constructs, the deposits often occur within the exclusive economic zone (EEZ) of these countries. Similar in general composition to polymetallic nodules, cobalt crusts began attracting investment in exploration for their easier accessibility and their higher cobalt percentage (up to 2 %), platinum (0.0001 %) and Rare Earth Elements (REE) in addition to the nickel and manganese they share in common with the nodules (International Seabed Authority 2015; United Nations 2013; cf. asteroid mineralogy classified by Kargel 1994).

The extraction of REE's from ferromanganese crusts involves complex mining and smelting technologies that far exceed the cruder methods envisaged in the 1970 and 1980s (and largely since unchanged) for nodules and sulphides. Both Japan and China are rapidly perfecting these technologies however, and in 2014 both countries signed 15 year exploration contracts with the Authority (Sunday Times 2013).

> Contractors are awarded licenses to explore specified parts of the deep oceans outside of their own national jurisdiction. Under the Authority's regulations, each contractor has the exclusive right to explore an initial area of up to 150,000 km^2. Over the first eight years of the contract,

half of this area is to be relinquished. In addition, each contractor is required to propose a programme for the training of nationals of developing States—a more practicable realisation of the Law of the Sea's initial aspiration to create a central pool of profits for general wealth re-distribution. Each contractor is also required to submit an annual report on its programme of activities, including environmental and scientific engagements, and prior to the commencement of its programme of activities under the contract, each contractor is also required to submit to the Secretary-General a contingency plan to respond effectively to incidents arising from its activities in the exploration area (International Seabed Authority 2015).

The cross-section of private and governmental contractor teams now involved is impressively diverse, and despite the U.S.' persistent failure to ratify the 1982 Convention on the Law of the Sea (and almost any other international treaty in the throes of its stymied domestic politics), U.S. institutions like Duke University are increasingly finding means of involvement through research if not direct exploitation. These teams include UK Seabed Resources Ltd, with national sponsorship from the UK Government (contract start 2013) and similar private-public alliances in Germany, Belgium, and France as well as Russia, China, India, Japan and the Republic of Korea as well as many smaller Pacific nations (International Seabed Authority 2015; BBC News 2014).

There is effectively a rush of activity to apply private-company exploration, mineralogy and mining talents to government-sponsored exploration and extraction licenses, at the price of training scientists and engineers from developing nations, particularly in relevant environmental sciences, and abiding by biodiversity and other oceans-protecting constraints. It's a mix that works. It is neither the single lever of world wealth re-distribution and social justice as originally envisaged by U. S. President Johnson in the 1960s, nor is it a whoever-gets-there first free-for-all leaving environment, science and ethics to be sorted by others in the aftermath. It is a remarkably nimble and unheralded achievement whose time has come for application to strikingly similar challenges in the exploration and exploitation of Outer Space (Bernasconi and Bernasconi 2004).

12.3 The Counter-Trend Toward Unregulated Competition in Space Is Accelerating, Just When Regulation Is Becoming Most Critical

Nine years old when the first pair of Apollo astronauts walked on the Moon, I enjoyed the thrill of this wonder-filled, live-televised field trip as perhaps only a schoolchild can, with complete faith in its promise of a shared future in space for all humanity, including of course for myself. And as a schoolchild I took on faith too, the expressed meaning and purposes of our visit there. I especially treasured a particular piece of promotional material produced and sponsored by NASA and the Gulf Oil Company, a book my father got with a tank of petrol just after that first landing in 1969.

Entitled *We Came in Peace*, it became my many-times read companion throughout all the remaining Apollo flights (Smith 1969). Along the way it regularly reinforced for me the popular space fictions of the time: that the Soviets had not really been interested in any race to the Moon, and if they were, we would like to go with them, not beat them; that space would not be used for weapons, but rather for science; and that a burgeoning program of exploration would have room for thousands like me:

> You can have an exciting and rewarding career in your country's program to explore and use space. When you finish college, many thousands of qualified young men and women will be needed. Before that time, the United States should have a large space station circling in orbit far above the earth. As many as a hundred people may be living on the station, carrying out scientific work. At the same time, the United States will probably have built a scientific station on the Moon. Astronauts will be making long trips on the Moon's surface to learn about everything there. Even more exciting, it is likely that astronauts in much larger spaceships than we have at present will be training to go millions of miles from Earth to Mars and Venus. Plans will be underway to reach ever more distant planets in our solar system (Smith 1969).

When I did graduate from MIT in 1982 of course the face of the space programme had utterly changed, and the underpinnings driving some of the over-expectations have been well and entertainingly examined (Burrows 1998; McCurdy 2011).

What the world gained in space exploration in those intervening years, however, brings another kind of wonder and inspiration. Remote sensing by satellite and interplanetary robotic spacecraft unveiled the detailed terrain of our solar system, giving us for the first time a view of the cloud-shrouded volcanic fields of Venus, the canyons and mountains of Mars, and the turbulent magma ocean of Jupiter's moon Io. Space-borne observatories revealed much of the structure, nature and extent of the universe, and now promise to help us explore a diverse menagerie of other stars' planetary systems.

In the decade ahead it is possible that the artefacts of humankind will resolve enough of our neighbouring solar systems, coupled with sufficient knowledge of our world and our solar system, to chart the likely frequency of water and life-bearing planets. We will make better and stronger guesses about how inevitable or incidental intelligent life may be, where there is life. The author's own planetary sciences work builds on hypotheses (Arkhipov 1998; Pinault and Crawford 2015) that we may at least find the merest traces of dust-sized alien manufacture, remnants of long-ago space-faring activities elsewhere in our galaxy, that may have deposited on our Moon, Mars, the asteroids and other comparatively accessible parts of our solar system. Through some combination of these possible means and discoveries we may identify and seek out our first candidate systems for contact and communication. What awaits us at one end of the possibilities could be a bonanza of the deepest knowledge and insight, a tapping into the lives, lore and wisdom of civilizations spanning the galaxy, or at the other a sobering confirmation of our aloneness, and an awareness that if there is to be even one civilization that touches the face of the galaxy it falls to us to make use of the fragile and precious opportunity to step out of our cradle.

This is the promise of space. But in our time competing uses for space are fast emerging, ones that threaten to hinder and thwart scientific understanding, bring damage to an ethical and aesthetic appreciation of the near-Earth space environment, and engender a destructive competition for energy and material resources in space. At best, these developments may confine the full benefits of space exploration to a powerful few. At worst, they will spawn conflicts that will make our nearby solar system an arena of destruction, and bring potentially fatal harm to Earth and humankind, and thereby, perhaps, the future of sentient life in the galaxy.

It was with some considerable broadmindedness that the creators of international treaty law for space in the 1960s restricted the use and exploration of space to peaceful purposes, and for the benefit all of all humanity. There was considerable pressure at the time to maximize the potential for seeking military advantage in outer space, and to plan for brute force use of manned colony vessels to expropriate the territory and resources of the nearby solar system. With great good fortune for us as a species, the Cold War rivalry between the space powers of the time, and the need those powers perceived to appeal to the have-nots of our planet, resulted instead in a healthy degree of restraint that channelled the space powers' energies into mainly scientific and non-destructive endeavours in outer space (Burrows 1998).

This may have been a temporary reprieve. These restraints seem fast eroding with the end of the Cold War balance of power and the emergence of many new space actors. Where the first United Nations plenary session on the uses of outer space, UNISPACE I in 1967, focused on resolving the Cold War rivalry, and UNISPACE II in 1982 dealt largely with developing-nation concerns about space exploitation, UNISPACE III in 1999 saw the beginning of the integration of commercial interests into international space law and policy, and now sixteen years later with no UNISPACE IV in sight, those interests increasingly own the debate.

In the United States, President Bill Clinton's signing of the Commercial Space Act of 1998 (NASA 2015) was the watershed moment that ultimately enabled the innovations of SpaceX and other new players; it also however opened doors to broad re-interpretations of the ownership of space scientific mineral data and celestial bodies themselves, with a focus on asteroids. Scientists were positioned to negotiate data from commercial providers and international research funds, and the Act opened a still-unfolding and often acrimonious debate on the ownership of Earth remote sensing data.

The Bill's main body was devoted to largely innocuous and widely lauded goals —privatizing NASA's routine launch operations, facilitating key contracts for the assembly and commercial use of the international space station, and the like (Bush 2002). Somewhat lost in the deeper detail of the Bill, and what back in 1998 seemed an issue for far-future consideration, Section 105 (c) classified 'space science data concerning the elemental and mineralogical resources of the Moon, asteroids, planets and their moons, and comets' as a 'commercial item.' It follows from this construction of the law that such data becomes the property of 'commercial providers,' to sell when and if they please, and at the price of their choosing. There are virtually no controls on the commercial sector's expropriation of this core product of the U.S. space programme. Though the Bill guaranteed NASA itself an option to buy space science data, and the U.S. State Department was required to somehow see to it that 'as part of overseas aid,' developing countries would get money to buy space science data too, knowledge about space would in future be priced for whatever the market could bear.

This commercialising of science data opened the way for would-be private space exploiters to make the case for physical claim of the space objects themselves. Start-up company SpaceDev promptly declared intent to claim ownership, science data and mining rights for any asteroid it might meet with its planned Near-Earth Asteroid Prospector. SpaceDev proudly proclaimed that an 'objective of the NEAP mission is to make an ownership claim of the asteroid on which NEAP lands,' which could 'set a precedent for private property rights in space,' which they also remarkably construed as a 'benefit to all of humankind by making space more accessible' (Landesmann 1998).

While SpaceDev would go on to other plans and projects before finally being acquired by the Sierra Nevada Corporation in 2008, its president Jim Benson had the financial market access and technical wherewithal to catch NASA's attention and garner support for its acquisitive goals from then-administrator Dan ('faster, cheaper, better') Goldin. In an interview of the time Goldin made international space law seem but an interfering bramble which warriors of space commerce like Jim Benson must penetrate. He related that 'there's a fuzzy area as to who owns what. I don't see it as greed—I see it as a man going for it, a real possibility for our progeny… Jim Benson is necessary for progress' (Landesmann 1998).

While the Benson-Goldin braggadocio and rhetoric has eased, today the mantle for asteroid development is squarely worn by Peter Diamandis' Planetary Resources Inc., and there is an ASTEROIDS Act, a direct descendent of the Commercial Space Act of 1998, now receiving renewed attention in the newly Republican-led U.S. Congress (Planetary Resources 2015; Tronchetti 2014). In parallel, the American unilateral approach to commercial space development seems likely to be matched by any number of other nation's space interests, and the makings of a free-for-all are in sight.

Perhaps more worryingly, the 1998 Act set in motion a chain of events with the U.S. Department of Defense that today is entwined in NASA's current proposals to capture an asteroid and move it into lunar orbit. At the time the Department of Defense seemed clearly eager to develop and provide for enforcement and protection of commercial interests in space. Gen. Thomas S. Moorman Jr, then vice chief of staff, United States Air Force, noted when the Commercial Space Act was first proposed that 'about twenty countries are expected to have space-based capabilities by 2000 and many others will purchase space derived information.' With apparent appetite he emphasized that 'as the importance of space grows and other countries begin to exploit the advantages of operating in the high ground, the next trend—controlling space—will become as important as controlling the seas or air. Eventually, an ability to protect, deny, disrupt, degrade and destroy space assets must be pursued if the United States and its allies want to ensure freedom of access and action in space—for example, the readily available commercially derived imagery ' (Moorman 1999).

Benson's, Goldin's, and Moorman's ambitions were underpinned by a broadly-applied logic. 'Commercial' space enterprise of the kind they endorsed is neither non-governmental nor market-driven. It instead concentrates a diversity of the government's research and development monies into the hands of a few industrial beneficiaries. This is the logic of what built the railroads and other crucial infrastructure in the early days of America's industrialisation, and with SpaceX's novel developments and achievements it may prove a game-changer again. But in applying these principles to space science data, whereby NASA would pay the research scientists, the scientists pay the 'private' entrepreneurs, and the Defense Department spends more government monies to protect them, in the name of bootstrapping a new industry, the actual benefits and sustainability of the business model become questionable, particularly in the face of unilateral action in a world newly full of space competitors, and against the backdrop of not-yet-abandoned aspirations to international cooperation and development.

I offer that there is a more balanced approach which is firmly grounded in equitable principles of international law, and yet still facilitative of a market-driven human destiny in space.

12.4 Distinguishing the Resources of Space and the High Seas Underlines the Even Greater Importance of an International Space Authority

Based on both industry and disadvantaged States' experiences with mining prospects for the deep seabed—some three decades of debate and negotiation before high-seas licenses were granted—mining in outer space might seem an issue to be confronted at some comfortable distance in the future. Indeed, prospects for extraterrestrial mining of the most accessible mineral resources on the Moon, Mars, or near-Earth asteroids, for delivery of materials to Earth may well be non-existent (Crawford 2015). All but the most dire scarcity or technical incapacity on Earth would make the expense of bringing such resources in any form to Earth wholly unattractive. And by the time such scarcity has arisen, one might hope that the industrial processing of minerals has become strictly an off-Earth activity, in respect to the needs of Earth's naturally life-sustaining environment.

But the mineralogical resources of the Moon, Mars and the asteroids differ from that of prospects for the deep seabed in at least one crucial respect. With habitable land never more than a few days' or weeks' journey away, deep seabed mining itself is not crucial to exploring, living and operating on (or under) the high seas. But the mining and processing of extraterrestrial materials in space will be essential to the establishment of Earth life and communities in extraterrestrial space. Even in Earth orbit, no practical, large scale structures for habitation or energy generation can be built without material taken from the relatively low gravity wells of the Moon or the near-Earth asteroids. Once in the gravity wells of these bodies, material extracted from them will be essential to providing water, living atmosphere, nutrients, and protective shielding from hard radiation (Elvis 2012; Lewis and Hutson 1993; Lewis 1996).

A key issue in thinking about lunar settlements for science is how to protect the Moon's environment to help reveal scientific information about its pristine state. Before even invoking an aesthetic argument, this is understood to be a first principle for environmental protection by at least one international lunar research group, and has become the basis of one of the first recognised 'environmental impact statements' for a celestial body (Foing 2007).

But the very first outposts for extended habitation of the Moon or Mars will require the mining and processing of extraterrestrial materials for their survival, and therefore at least some major disturbance of their surface is likely to ensue.

Given that extended, hard-fought negotiations have already produced an international Authority and Enterprise system for the deep seabed, then, it would not seem premature to structure a system for extraterrestrial mining that would also accommodate the competing interests of industry, disadvantaged states, science, and the environment in outer space. To not create such an international system for outer space would be like States and individuals preparing to exploit and occupy the international deep sea bed without any reference to a system of laws and agreed principles.

12.5 Principles of Outer Space Law Still Bear Application Beyond the Moon Treaty

The 1967 Outer Space Treaty remains the touchstone of the international public law of outer space, setting the norms by which all other related treaties and agreements must be judged. Four other major space treaties have since followed, building on the standards set by the 1967 Treaty.

The Treaty on Principles Governing the Activities of States in the Exploration and Use of Outer Space, Including the Moon and Other Celestial Bodies (the '1967 Space Treaty') sets out to regulate outer space activities and establish key parameters for space exploration. The Treaty presently has 102 state parties, and the vast majority have both signed and ratified, including the United States. G. Zhukov and Y. Kolosov of the former U.S.S.R. were instrumental in developing many of the fundamental principles of this and subsequent treaties, and highlight the following as key principles of the 1967 Outer Space Treaty (Zhukov and Kolosov 1984):

- Freedom of exploration and use of outer space and celestial bodies;
- Non-appropriation of outer space or celestial bodies;
- Exploration and use of outer space and celestial bodies in accordance with the fundamental principles of international law, including the basic principles of the UN Charter;
- Demilitarization of outer space and celestial bodies;
- Retention by states of sovereign rights over space objects launched;
- International responsibility of states for national activities in space, including liability for damage caused by space objects;

- Prevention of potentially harmful consequences of experiments in outer space and on celestial bodies;
- Assistance to personnel of spacecraft in the event of accident, distress, or emergency landing; and
- International cooperation in the peaceful exploration and use of outer space and celestial bodies.

The preamble of the Treaty encapsulates its foremost guiding principle: 'the common interest of all mankind in the progress of exploration and use of outer space… should be carried on for the benefit of all peoples irrespective of the degree of their economic development.' States cannot expropriate portions of outer space or celestial bodies and claim such as within their national sovereignty. And under their prescribed responsibilities, they cannot sanction or recognize claims to such expropriation by private entities or individuals. States do control and retain jurisdiction over objects and personnel sent to outer space, 'including objects landed or constructed on a celestial body.'

The 1967 Outer Space Treaty borrowed concepts and wording from the Antarctic Treaty and the Nuclear Test Ban Treaty to frame its introduction and concluding sections, and otherwise built from prior UN resolutions. The Antarctic Treaty's text contributed to the 'Space Treaty's important substantive provisions' and the Nuclear Test Ban Treaty 'contributed to its formal and final clauses.' The overarching achievement of all these treaties was to 'avoid the extension of present national rivalries into this new field' (Stovitz and Loomis 1985).

Other Treaties quickly built from the premises of the 1967 Outer Space Treaty and are well enshrined in operating practice and case law:

- The Agreement on the Rescue of Astronauts, the Return of Astronauts and Objects Launched into Outer Space (1968)
- The Convention on International Liability for Damage Caused by Space Objects (1972)
- The Convention on Registration of Objects Launched into Outer Space (1975)

Next in line however, the Moon Treaty of 1979, has all but foundered as a functioning protocol. The Agreement Governing the Activities of the States on the Moon and other Celestial Bodies marked the beginning of a new era, 'which extended beyond the 1967 Principles Treaty [1967 Space Treaty]' It represented 'the divisions between space-resource states and earth-bound states and between Soviet and American ideologies [which were becoming] increasingly deep as the value of space became more evident to all nations' (Goldman 1985, 1996).

Its sixteen ratifying parties, none of whom have self-launching capability for human spaceflight, affirm the Moon Treaty as a fundamental principle in international space law, but it is widely seen to have over-reached, putting principles of comparatively generous wealth re-distribution far ahead of commercial incentives. It has thereby unfortunately taken down with it some sorely lacking environmental and scientific guiding principles (Baca 1993). The Moon Treaty 'refines some of the principles set forth in the [1967] Space Treaty' by 'prohibit[ing] military uses of the moon, require[ing] awareness of environmental implications of activities, encourage[ing] disclosure of scientific research, and render[ing] assistance [in space] to those in need.' (Hoffstadt 1994). Many questions remained unanswered as to exactly how their interests would be protected as the space powers began exploring and exploiting the Moon and other celestial bodies. For this reason, proponents of the Moon Treaty asserted it was essential because it specifically addressed the Moon and reflected in crystalline language the degree of states' interests in the resources of the Moon. Most commentators were dismissive of both the Moon Treaty and its context:

> The Moon Treaty represented the views of a then-influential group of developing-state economists who viewed Third World economic problems as resulting from Western exploitation. This group's push for a 'New International Economic Order' was reflected in the Moon Treaty's hostile treatment of not only national sovereignty (already forbidden by the 1967 Treaty), but also of private property rights—with any for-profit exploitation of space resources to be undertaken only by a monopolistic international organization that would ensure that a share of the profits went to developing states (Reynolds and Merges 1989).

Other commentators put a more optimistic face on the group of space law treaties overall:

> Collectively, [the international space treaties] provide that space shall be free for all mankind to use in a peaceful manner. The theory behind the treaties is that all of mankind should benefit equally from what is found in outer space. The treaties were perhaps one of the first real attempts at establishing a global community that would work together to accomplish a goal. Space would not be divided up, as were the land masses on earth, through conquest and colonialism. Rather, the vision for space was one of humans working in harmony to better the lives of all mankind by exploring and possibly exploiting space resources for the good of all, in the spirit of cooperation and harmony (Keefe 1995).

United States ambassador to the United Nations Arthur Goldberg, in seeking to secure ratification of the Outer Space Treaty in 1967 by the U.S. Senate, made assurances that continue to mark American politicians' views of its international obligations in space.

> Senators wondered if the United States wanted to make these principles binding. What was meant, for instance, by 'for the benefit and in the interests of all countries'? Did this mean, asked Senator Gore [Sr.], that the United States was obliged to make outer space available to all? Not quite, said Goldberg, it merely stated a goal subject to further refinement… one concludes that the phrase did not oblige parties to the treaty to share their technology or its fruits with others: no 'international socialism' in space.
> …Chairman Fulbright clarified that while all weapons were prohibited on the celestial bodies, only weapons of mass destruction were banned from earth orbit. [Senator Gore Sr. raised] 'space for all mankind.'….Goldberg tersely explained that space as the 'province of all mankind' was meant sort of 'freedom of the seas' clause, no more.
> Thus the UN Outer Space Treaty of 1967, ratified by a vote of 88 to 0 … denuclearized outer space and demilitarized the moon. But it did not demilitarize outer space. As for space being 'for the benefit of all peoples irrespective of the degree of their economic and scientific development,' the [U.S.] negotiators described it as a vague principle with no foreseeable application. In terms of the 'space for peace' globalists, therefore, the space treaty was all show and little substance (McDougall 1985).

With NASA's phase-shift toward commercialisation of scientific data under the 1998 Commercial Space Act, the 'show over substance' powers of the outer space treaties became all the clearer. In a report on International Cooperation in Space Activities for Enhancing Security in the Post-Cold War Era, the UN Secretary General emphasized that 'commercialization of certain space technologies, particularly remote sensing imagery, has resulted in high costs which many developing countries cannot afford. Those high costs are limiting developing country access to vital information just when commercialization suggests it will become available to them (UN Office for Outer Space Affairs 1999).

'A less visible, but equally problematic, effect of commercialization,' the Secretary General continues, 'is the tendency of developing countries to invest scarce resources in space systems and technologies that are operated primarily by foreign consultants or contractors, rather than investing in the education and training that is necessary to develop local expertise and competence in the use of space systems and technologies.' The Commercial Space Act exacerbated these inequities by specifically singling out remote sensing data as a commercial item, while providing hopeful but weak wording directing the State Department to fill in the gap by some unspecified process of allocating international aid for the data's purchase.

12.6 Pertinent Applications of Sea-Use Law to Outer Space Law Can Provide a Critical Framework and Required Enforcement Mechanisms

Under the Outer Space Treaty, asteroids and other bodies of our solar system are subject to international law. But in addition, specifically with respect to working out the nature and meaning of our common heritage, principles of customary international law, the 1994 Law of the Sea Convention and its terms also apply. Its key premises are that the oceans and their resources comprise the common heritage of all peoples, and that in respect of this, all parties to the Convention must agree to all its elements without reservation, and not merely choose those parts that might best benefit a particular interest (Wolfrum 2010).

The Law of the Sea Convention creates crucial precedent for the directly analogous issues of balancing international interests in science, environment, resources and security in the commonly shared realm of outer space. The terms were negotiated in detail by all the world's nations, large and small, seafaring and landlocked, for more than a decade. Officially entering into force in 1994, the result is a highly robust agreement that embodies a 'package deal' of hard-won compromises ranging from the navigation rights of warships, to fishing and mining on the high seas, to the conduct of scientific research and the protection of the marine environment (Van Dyke 1988; Oxman 1997, 2006).

Once the Convention's offspring, the International Seabed Authority, opened its doors in Jamaica in 1996, the U.S. while not ratifying its mother Treaty, after much intense negotiation joined with the majority of the world's nations in mandating the Authority to weigh and approve exploration rights for seabed mining on the high seas, with due attention to scientific and environmental concerns, stipulating that through the Authority mining investors must partner with, and make mining technology available to, companies from developing nations.

The 1994 Law of the Sea Convention and its operating arm for commercial resource development, the International Seabed Authority, thereby provides the strongest legal and regulatory analogies for outer space. The U.S. weighed into support the International Seabed Authority precisely to prevent its becoming a more-empowered version of the Moon Treaty, but in doing so has helped anchor a robust working institution, ironically with European and other Asian-Pacific commercial-national partners becoming the first and main beneficiaries. In this manner the Law of the Sea Convention and the International Seabed Authority bear more immediate relevance than the 1967 Outer Space Treaty and the work of the UN Committee on Peaceful Uses of Outer Space, because they have given life to detailed principles and working entities, and are empowered by hard-won worldwide consensus.

Thus the Law of the Sea Convention provides not only governing precedent, but in the International Seabed Authority, a working model for international activities in space. An analogous International Space Authority would not block the development and commercialization of space, but rather allow it to proceed in an internationally equitable fashion with due regard to scientific and environmental interests, while providing an agreed framework and mechanism for enforcement.

12.7 Recognising that Human Habitable Space Is not a Frontier Further Underscores the Need for an International Space Authority

Even if as a species we choose to ignore lessons learnt from the environmental damage we've inflicted on our home planet, and we are prepared to exploit what we may perceive to be a new and untrammelled frontier as we please once again, for reasons of science alone we should give pause. As a scientific resource our solar system is a unique environment requiring careful progress. However many others there are, this is the only planetary system that can shed light on how life on Earth began, and it is possible that we are the only bearers of sentient life. The ultra-thin atmosphere of the Moon and the billions of years of galactic history captured on its surface, the largely untouched potential surficial biosphere of Mars and its pristine potential life-bearing habitats deep below the surface, the insights afforded by the Rosetta mission and its first landing on one of our comets, and any new asteroid encounters and sample return missions, all offer precious and delicate clues as to how we came about, and what we might look for in other systems.

Even with the excitement of Kepler's many discoveries of potentially life-bearing planets, the nearest planetary systems are very, very far away, and our solar system will likely be our best origins laboratory for generations to come (Lammer and Kodachenko 2014). Certainly Jupiter's moon Europa, with its tantalizing tidal energy sources, deep-water ocean circulating under kilometres of protective ice, and hints of organic compounds, invites a carefully considered and scientific, rather than commercial, approach —for there we may find not mere clues, but life itself (Toon 1997; Taylor 1998).

With eight planets, hundreds of moons, and many thousands of fair-sized asteroids to hand, commercial space proponents might be forgiven for espousing a frontier ethic premised on unlimited and limitlessly exploitable bounty. We are after all but a speck in a likely myriad of other planetary systems, and perhaps predictably the notion of property rights, as opposed to species, societal, scientific, and environmental trusts and obligations, has received considerable attention (Simberg 2012a, b; Pop 2001, 2012; Collins 2008; Sterns and Tennen 2002; Reinstein 1999).

But in terms of pleasingly preserved, habitable environments, our solar system offers limited choices indeed. The gas giants and their moons swathed in lethal radiative and magnetic energies, Mercury in its blazing proximity to the Sun, Venus in the death grip of greenhouse gases and far icy Kuiper Belt objects do not invite immediate settlement. Homesteaders will be largely limited to Mars, our own Moon, and some of the largest asteroids. Taken together they barely offer the same surface area for habitation and development as the Earth itself: we likely have just one more Earth to go, before we have to start building new ones. How we develop our societies for what may be this only other foreseeable option is an important study in itself, and is treated with care and imagination elsewhere in this volume (see too Cockell 2013; Hartmann 1986).

Will we sense and act on a special consideration for the grand canyons and volcanoes of Mars, the historical features of the Moon, and feel a preservationist appreciation of the largest asteroids, before we overwhelm them? That human habitability should itself be a standard for our disposition of the space environment is itself an idea worthy of deep examination, one that philosopher Holms Rolston III used to suggest a thoughtful set of rules for probing toward a solar system exploration ethic that seems to me at least both pertinent and attractive (Rolston 1986):

(1) *Respect any natural place spontaneously worthy of a proper name.* By this Rolston refers to places where there is enough particularity, differentiation, and integration of locus, enough provinciality of identity to call for protection. He notes such a rule would not likely call for a human duty to every crater on the Moon, but maybe to some of them, or representative types. Neither does he ascribe a particular value, or definition of protection. Rather, he suggests this rule to begin thinking about how much such a place might count, given that it 'is.'

(2) *Respect exotic extremes in natural projects.* Other bodies in the solar system, Rolston notes, will see nature giving expression to potentials that cannot be realized on Earth.

In wildness and magnificence unbounded by earthly constraints, Rolston posits a considerability beyond human utility.

(3) *Respect places of historical value.* Humans ought to preserve those places that have been more eventful than others, Rolston offers.

(4) *Respect places of active and potential creativity.* This rule captures Rolston's determination to warn of the tragedy that might be extended in taking habits that have reduced speciation on Earth to the destruction of new generative capabilities elsewhere in the solar system in future.

(5) *Respect places of aesthetic value.* Some planets, moons, asteroids and comets will have more symmetry, harmony, elegance, beauty, and grandeur than others. Rolston suggests that such high order considerations should have weight in counting the value of preservation. Other values might be complexity, fertility, rarity, information content, historical significance, potential for development, and stability.

(6) *Respect places of transformative value.* By this Rolston means to see preserved those places that radically transform human (and later, perhaps other species') perspective. The solar system itself as a place that will transform the earthbound value system may be this rule's most profound expression.

Even in the perhaps unlikely event that our solar system should prove innately lifeless beyond Earth orbit, Rolston's framework underscores that human aesthetics alone, and humanity's own powers of future potential, together with those of other Earth species, will bring complex wonders and diverse values and perceptions requiring equitable means and principles to share in their richness.

12.8 Applying an International Seabed Authority Model to an International Space Authority Yields Multiple Benefits

Equities in common heritage are well addressed in the working model of the International Seabed Authority, and would well be adapted to an International Space Authority. Commercial investors are by no means prohibited from mineral and energy exploration and development, but rather are given a means of reconciling and enforcing competing claims.

Requirements to partner with companies from disadvantaged nations are not onerous and go a long way toward better peace and security, and hence lower insurance and enforcement costs for development. No technology give-aways are entailed—rather, primary investors must make technologies available for purchase at reasonable market rates. All that is required is working through a central authority for development licensing, a degree of profit sharing with disadvantaged partners that is likely more than made up in reduction of other costs, and due regard for scientific and environmental concerns specified by the Authority under the relevant Convention.

There are numerous examples of commercial, scientific and environmental tradeoffs would benefit from an International Space Authority, much as

Internet Corporation for Assigned Names and Numbers has done for the internet domain on Earth. Radio-wave astronomy for example allows us to 'listen' to tremendous quantities of highly revealing information about stars, galaxies, and other energetic bodies in our universe; one day we may even be privileged to listen into the transmissions of other civilizations at these frequencies, perhaps from the earth-like planets detected from large telescope arrays whose construction might themselves be best facilitated by an International Space Authority. The ideal place for listening is the far side of the Moon, which because it always faces away from Earth, is one of the few places in the solar system entirely free of earth's prodigious radio noise. But without the regulation and enforcement of an International Space Authority or the like to protect it, the lunar far side will no doubt be awash in its own radio noise as space commercialisation proceeds apace.

An International Space Authority can also do much to avoid the helter-skelter weaponisation of space. Each space buccaneer will not feel obligated to carry its own cannon if there is a space navy under international agreement to protect it. One of the greatest champions of the Law of the Sea Convention, centralized Seabed Authority and all, is the U.S. Navy. The Navy appreciates that developing country concessions, allowing key navigation rights for all, helps create equitable stability and greater security. In exchange the Navy does not begrudge disadvantaged nations' interests in sharing the common resources of the international seabed, and indeed can be called upon to protect those rights.

An unconsidered implementation of the 1998 Commercial Space Act and its successors including the present-day proposed ASTEROIDS Act can engender policy promoting the weaponisation of space. As well captured in General Moorman's comments, the U.S. military would seem well prepared to step into defend any commercial interests in space. But even purely commercial endeavours raise weapon hazards. Industrial manipulation of asteroids for mining, for example, raises the prospect of asteroid material being directed at Earth for more nefarious purposes. As Carl Sagan warned any manoeuvring of asteroids, whether for mining purposes or for so-called planetary defence, raises the spectre of their use as a powerful, stealthy, and unstoppable weapon of mass destruction (Sagan 1994). NASA is presently mulling whether to develop exactly these capabilities as part of its experience-building in advance of human exploration of Mars[2]—All the more reason the institution of an International Space Authority to oversee both asteroid mining licensing in the context of all relevant environmental impacts (!) as well planetary defence, and the shared infrastructure and systems to support them, would best stabilise commercial opportunity, provide equitable development on behalf of all humankind, and best protect us from asteroid and comet hazards while discouraging their appropriation as weapons.

[2]It remains to this author almost irresistible to think that a return to the Moon and/or a direct path to Mars would reap more benefits.

12.9 From International Space Authority to World Space Agency?

To all who suspect that behind every international agreement is the path to world socialism, the idea of a world space agency is likely pure anathema, and might seem to have best deserved a death alongside the moribund Moon Treaty, if not with the collapse of the Berlin Wall. But with the complexity of ever larger projects in space required to conduct the biggest of Big Science, and the fast-expanding range of transnational commercial interests and talents constantly crossing borders, the benefits of pooling resources make it now timely to put this discussion back on the world agenda. A natural outgrowth of an International Space Authority that balances, regulates and enforces competing scientific, environmental and commercial interests could become the backbone for shared investments at the grandest human scale.

As with the International Seabed Authority, the UN itself, through an International Space Authority, would seem a natural organization for monitoring and enforcement of regulated activities in outer space. A succession of Secretary Generals have recommended and requested that the United Nations be equipped with its own earth remote sensing satellites to monitor climate hazards, crop and environmental threats, refugee and illegal troop movements, forced labour camp abuses and the like. These recommendations now seem particularly sensible in light of national legislation like the U.S. Commercial Space Act, which limits U. N. access to such data. Under the Act's regime, much remote sensing data will be routinely withheld under U.S. security requirements, and that which is made available will have to be either bought from commercial providers, or negotiated from the U.S. government for purchase on their behalf. Both burdensome and untimely, this system eviscerates any UN response capability based on satellite remote sensing.

Thus earth remote sensing might be a logical start for a UN World Space Agency, and an immediate counterbalance for the destabilizing effects of the U.S. Commercial Space Act and national legislation it seems likely to encourage in fast-emerging space powers like China. Given that remote sensing data is also notoriously subject to manipulation, greater trust in the truth and accuracy of what the satellites reveal would likely be accorded the UN, particularly where UN security actions are to be invoked.

A United Nations agency for space is not a new idea. Crawford (1981) notably offered prescient insights. On the back of the International Geophysical Year Antarctica was proposed as a direct analogue for controls on Outer Space (Jessup and Taubenfeld 1959). In 1958 the U.S. Library of Congress completed a study under congressional direction, proposing a UN space agency to be charged with all matters relating to the exploration of space. At that time, both the United States and the Soviet Union put forward proposals for United Nations management of outer space.

> [In 1958] Senator Jackson addressed the UN General Assembly to demonstrate the unanimity of American opinion behind 'space for peace.' He asked, among other things, that a UN space committee 'consider the future form of internal organization in the UN which would best facilitate cooperation in this field.

> Such language could easily be interpreted as an invitation to the UN to assume strict management of all human activity in space. Certainly no enterprise fell more clearly under UN jurisdiction. The Soviets' own resolution called for a ban on all military uses of space, elimination of foreign bases, international control of space, and a UN agency to include an international program for launching long-range rockets (Shepard 1958).

This new organization could be charged with asserting the core principles of the Outer Space Treaty, ensuring the lasting demilitarization of space, and the rejection of all claims to space objects and data not sanctioned by the UN organisation. To comply with international law and the principles of the UN Charter, all key decisions of the UN International Space Agency could be put to vote for ratification by the General Assembly. The organization could exist as a semi-independent body, alongside the International Space Authority, to best accommodate the integration of private commercial interests with scientific and environmental concerns and requirements.

There is otherwise no international organisation which can cope with private claims made in space; no guidance as to the division of space with regard to scientific, environmental, or commercial needs, no comprehensive limitations on the weaponisation of space and no independent cooperative framework for joint projects that could involve representatives of all peoples. Legislative and enforcement mechanisms would have to be defined, but with both the World Court and the International Criminal Court now enjoying full powers and relative independence under international law and UN auspices, a good working foundation for legal and enforcement mechanisms is already emerging.

Many of the building blocks are already in place. Many space agency representatives from around the world already contribute to the negotiation and advance of a Global Roadmap for Space Exploration. The International Seabed Authority is already partnering nations and companies small and large in mining exploration, science and training. The UN continues to enjoy a hard-won reputation for prestigious service in the interests of advancing humanity's best interests and potential —how much more so if an ambitious student from anywhere on the planet could aspire to work for the UN World Space Agency?

12.10 Author Biography

Dr Lewis Pinault is a researcher at the UCL/Birkbeck Centre for Planetary Sciences, Senior Innovation Fellow at Imperial College London, and Space Innovations Advisor at the UK Government sponsored Satellite Applications Catapult. Following his graduation at MIT in Political Science and Ocean Engineering, he worked with a major Japanese steel and shipbuilding conglomerate and piloted a subsea mining venture for cobalt encrustations in what is today the Hawaiian Exclusive Economic Zone. As a Fulbright Scholar at LSE he completed his MSc in Sea-Use Law and Technology and after a decade in consulting spanning The Boston Consulting Group and his partnership at Coopers&Lybrand (PwC), he became a NASA Space Grant

Fellow at the University of Hawaii, completing his Juris Doctorate in international public law in combination with planetary geophysics, where as a meteoriticist he focused on both asteroid origins and their mining potential. At that time he was sponsored by the UN Office of Outer Space Affairs and NASA to join the 1999 Third UN Conference on Peaceful Uses of Outer Space as a Space Generation delegate, where he successfully campaigned for the incorporation of an International Space Authority proposal into the Conference Report. In the ensuing decade and before resuming his research at the Centre for Planetary Sciences he served as an innovation advisor to large R&D organisations including at Royal Dutch Shell and the Airbus Group, as well as ESA's Industry Advisor Group. Currently he works on the Education and Science teams at the independently financed Lunar Mission One team, where he is Director of Lunar Missions Labs, targeting robotic core sample retrieval at the edge of the Moon's South Pole Aitkin Basin.

References

Anand, M., Crawford, I. A., Balat-Pichelin, M., Abanades, S., van Westrenen, W., Péraudeau, G., et al. (2012). A brief review of chemical and mineralogical resources on the Moon and likely initial In Situ Resource Utilization (ISRU) applications. *Planetary and Space Science, 74*, 42–48.

Arkhipov, A. V. (1998). Earth-Moon system as a collector of alien artefacts. *Journal of the British Interplanetary Society, 51*, 181–184.

Baca, K. A. (1993). Property rights in outer space. *Journal of Air Law and Commerce, 1041*, 1069.

Benaroya, H. (2010). *Turning dust to gold: Building a future on the Moon and Mars*. Chichester: Springer-Praxis.

Bernasconi, M. C., & Bernasconi, C. (2004). Why implementing the space option is necessary for society. *Acta Astronautica, 54*, 371–384.

Bowler, S. (2014). The new race to the Moon. *Astronomy and Geophysics, 55*, 5.15–5.17.

Burrows, W. (1998). *This new ocean*. New York: Random House.

Bush, L. (2002). International space station commercialization. *Policy Technology in Society, 24*, 69–75.

Chakhmouradian, A. R., & Wall, F. (2012). Rare earth elements: Minerals, mines, magnets (and more). *Elements, 8*, 333–340.

Cockell, C. S. (2013). *Extra-terrestrial liberty—an inquiry into the nature and causes of tyrannical government beyond the earth*. Edinburg: Shoving Leopard.

Collins, D. (2008). Efficient allocation of real property rights on the planet Mars. *Boston University Journal of Science and Technology Law, 14*, 201–219.

Crawford, I. A. (1981). On the formation of a global space agency. *Spaceflight, 23*, 316–317.

Crawford, I. A. (2006). The astrobiological case for renewed robotic and human exploration of the Moon. *International Journal of Astrobiology, 5*, 191–197.

Crawford, I. A. (2014). Science enabled by the global exploration roadmap. *Astronomy and Geophysics, 55*, 3.20–3.23.

Crawford, I. A. (2015). Lunar resources: A review. *Progress in Physical Geography, 39*, 137–167.

Crawford, I. A., & Joy, K. H. (2014). Lunar exploration: Opening a window into the history and evolution of the inner solar system. *Philosophical Transactions of the Royal Society, A372*, 20130315.

Crawford, I. A., Anand, M., Cockell, C. S., Falcke, H., Green, D. A., Jaumann, R., & Wieczorek, M. A. (2012). Back to the Moon: The scientific rationale for resuming lunar surface exploration. *Planetary and Space Science, 74*, 3–14.

Deep Space Industries (2015). deepspaceindustries.com.
Delgado-López, L. (2014). Beyond the Moon agreement: Norms of responsible behaviour for private sector activities on the Moon and celestial bodies. *Space Policy*. http://dx.doi.org/10.1016/j.spacepol.2014.09.003 (in press).
Duke, M. B., Gaddis, L. R., Taylor, G. J., Schmitt, H. H. (2006). Development of the Moon. In Jolliff, B. L., Wieczorek, M. A., Shearer, C. K., Neal, C. R. (eds) *New Views of the Moon*, Reviews in Mineralogy and Geochemistry (Vol. 60, pp. 597–656).
Ehrenfreund, P., et al. (2012). Toward a global space exploration program: A stepping stone approach. *Advances in Space Research, 49*, 2–48.
Ehrenfreund, P., Race, M., & Labdon, D. (2013). Responsible space exploration and use: Balancing stakeholder interests. *New Space, 1*, 60–72.
Ehricke, K. A. (1985). Lunar industrialization and settlement: Birth of polyglobal civilisation. In W. W. Mendell (Ed.), *Lunar bases and space activities of the 21st century* (pp. 827–855). Houston: Lunar and Planetary Institute.
Elvis, M. (2012). Let's mine asteroids for science and profit. *Nature, 485*, 549.
Foing, B. (2007). Personal communication, ESTEC.
Goldman, N. C. (1985). The Moon Treaty: Reflections on the proposed Moon treaty, space law, and the future. In J. E. Katz (Ed.), *People in Space* (pp. 140–150). New Brunswick: Transaction Books.
Goldman, N. C. (1996). *American space law: International and domestic*. San Diego: Univelt.
Hall, A., Barton, A., & Heward, A. (2013). Google lunar XPRIZE: New space entrepreneurs go back to the Moon—for good. *European Planetary Science Congress, 2013*, 1062.
Hartmann, W. K. (1985). The resource base in our solar system. In B. R. Finney & E. M. Jones (Eds.), *Interstellar migration and the human experience* (pp. 26–41). Berkeley: University of California Press.
Hartmann, W. K. (1986). Space exploration and environmental ethics. In E. C. Hargrove (Ed.), *Beyond spaceship earth: Environmental ethics and the solar system* (pp. 119–139). San Francisco: Sierra Club Books.
Heiken, G. H., Vaniman, D., & French, B. M. (Eds.). (1991). *The lunar sourcebook: A user's guide to the Moon*. Cambridge: Cambridge University Press.
Heim, B. E. (1990). Exploring the last frontiers for mineral resources: A comparison of international law regarding the deep seabed, outer space, and Antarctica. *Vanderbilt Journal of Transnational Law, 23*, 819, 832.
Hoffstadt, B. M. (1994). Moving the heavens: Lunar mining and the 'common heritage of mankind' in the Moon treaty. *UCLA Law Review, 575*, 583–585.
Inspiration Mars (2015). www.inspirationmars.org.
International Seabed Authority (2015). www.isa.org.jm/.
ISECG (International Space Exploration Coordination Group) (2013). The global exploration roadmap. http://www.globalspaceexploration.org/web/isecg/news/2013-08-20.
Jaumann, R., et al. (2012). Geology, geochemistry and geophysics of the Moon: Status and current understanding. *Planetary and Space Science, 74*, 15–41.
Jessup, P. C., & Taubenfeld, H. J. (1959). *Controls for outer space and the antarctica analogy*. New York: Columbia University Press.
Jolliff, B. L., Wieczorek, M. A., Shearer, C. K. & Neal, C. R. (Eds.) (2006). *New views of the Moon* (Vol. 60, pp. 1–721). Reviews in Mineralogy and Geochemistry.
Joy, K. H., Grindrod, P. M., Crawford, I. A., Lintott, C. T., Smith, A., Roberts, D., Fortson, L., Bamford, S., Cook, A. C. R., Bugiolacchi, Balme, M. R., Gay, P. (2009). Moon Zoo: Utilizing LROC lunar images for outreach and lunar science. Lunar Reconnaissance Orbiter Science Targeting Meeting. LPI contribution (Vol. 1483, pp. 64–65). doi.10.1111/j.1468-4004.2011.52210.
Kargel, J. S. (1994). Metalliferous asteroids as potential sources of precious metals. *Journal of Geophysical Research, 99*, 21129–21141.
Keefe, H. (1995). Making the final frontier feasible: A critical look at the current body of outer space law. *Santa Clara Computer and High Technology Law Journal, 11*, 345–346.

Lammer, H., & Kodachenko, C. (2014). Characterizing stellar and exoplanetary environments. Springer Astrophysics and Space Science Library.

Landesmann (1998). Asteroids for sale. *The New Yorker, 42*, 9 November, 1998.

Lewicki, C., Diamandis, P., Anderson, E., Voorhees, C., & Mycroft, F. (2013). Planetary resources —the asteroid mining company. *New Space, 1*, 105–108.

Lewis, J. S. (1996). *Mining the sky: Untold riches from the asteroids, comets and planets*. Reading, Massachusetts: Helix Books.

Lewis, J. S., & Hutson, M. L. (1993). Asteroidal resource opportunities suggested by meteorite data. In J. Lewis, M. S. Matthews, & M. L. Guerrieri (Eds.), *Resources of near earth space* (pp. 523–542). Tucson: Tucson University Press.

Lunar Mission One (2015). lunarmissionone.com.

Markussen, J. M. (1994). Deep seabed mining and the environment: Consequences, perceptions, and regulations. In: H. O. Bergesen & G. Parmann (Eds.), *Green globe yearbook of international co-operation on environment and development* (pp. 31–39). Oxford: Oxford University Press.

Mars One (2015). mars-one.com.

McCurdy, H. E. (2011). *Space and the American imagination* (2nd ed.). Baltimore: John Hopkins University Press.

McDougall, W. A. (1985). *The heavens and the earth: A political history of the space age*. New York: Basic Books.

McKay, C. P. (2013). The case for a NASA research base on the Moon. *New Space, 1*, 162–166.

Metzger, P. T., Muscatello, A., Mueller, R. P., & Mantovani, J. (2013). Affordable, rapid bootstrapping of space industry and solar system civilisation. *Journal of Aerospace Engineering, 26*, 18–29.

Moon Express (2015). http://www.moonexpress.com/.

Moorman, T. (1999). The explosion of commercial space and the implications for national security. *Air Power Journal, 8*(1), 6–31 (Spring 1999).

NASA (2015). *Commercial Space Act of 1998*. Retrieved January 25, 2015. http://www.nasa.gov/offices/ogc/commercial/CommercialSpaceActof1998.html.

BBC News (2014). *UK Seabed Resources Joins Deep Ocean Mineral Mining Rush*. Retrieved January 18, 2015 from https://ramumine.wordpress.com/2014/04/27/uk-seabed-resources-joins-deep-ocean-mineral-mining-rush-2/.

Oxman, B. H. (1997). Human rights and the United Nations convention on the law of the sea. *Columbia Journal of Transnational Law, 36*, 399–429.

Oxman, B. H. (2006). The territorial temptation: A siren song at sea. *American Journal of International Law, 100*, 830–851.

Pinault, L. J. (2000a). The international space authority and its role in lunar exploration. In *Proceedings on the Fourth International Conference on the Exploration and Utilisation of the Moon*, sci.esa.int/science-e/www/object/doc.cfm?fobjectid=39479.

Pinault, L. J. (2000b). Outer space resources: Power and justice in the making, use and ownership of space science data. Space Grant Report, University of Hawaii. www.spacegrant.hawaii.edu/reports/07_SP98-FA98/Pinault.pdf.

Pinault, L. J. & Crawford, I. A. (2015). Advancing a search for non-terrestrial artefacts on the Moon. *European Lunar Symposium*, 2015. els2015.arc.nasa.gov.

Planet Labs (2015). www.planet.com.

Planetary Resources (2015). http://www.planetaryresources.com/.

Pop, V. (2001). The men who sold the Moon: Science fiction or legal nonsense? *Space Policy, 2001*(17), 195–201.

Pop, V. (2012). The property status of lunar resources. In V. Badescu (Ed.), *Moon: Prospective energy and material resources* (pp. 553–564). Heidelberg: Springer.

Reinstein, E. J. (1999). Owning outer space. *Northwestern Journal of International Law & Business, 20*, 59–97.

Reynolds, G. H., & Merges, R. P. (1989). *Outer space: Problems of law and policy*. Oxford: Westview Press.

Rolston, H. (1986). The preservation of natural value in the solar system. In E. C. Hargrove (Ed.), *Beyond spaceship earth: Environmental ethics and the solar system* (pp. 140–182). San Francisco: Sierra Club Books.

Sagan, C. (1994). *Pale blue dot*. New York: Random House.

Shackleton Energy Company (2015). http://www.shackletonenergy.com/.

Shepard, M. (1958). An international outer space agency for peaceful purposes - a brief review of various proposals and an analysis of possibilities. Senate, *National Aeronautics and Space Act*, 85th Cong., 2nd. Sess., 1958, pt. 2:388.

Simberg, R. (2012a). Homesteading the final frontier: A practical proposal for securing property rights in space. *Competitive Enterprise Institute*, Issue Analysis 2012, No. 2. http://space-settlement-institute.org/support-files/space-homesteading-simbergreport.pdf.

Simberg, R. (2012b). Property rights in space. *The New Atlantis, 37*, 20–31.

Smith, Leroi. (1969). *We came in peace: The story of man in space*. Gulf Oil Company: Classic Press Inc.

Space Exploration Technologies (2015). spacex.com.

Space Flight 101 (2015). http://www.spaceflight101.com/china.html; http://www.spaceflight101.com/india.html.

Spudis, P. D. (2001). The case for renewed human exploration of the Moon. *Earth, Moon and Planets, 87*, 159–171.

Spudis, P. D., & Lavoie, A. R. (2011). Using the resources of the Moon to create a permanent cislunar space faring system. *Proceedings of the American Institute of Aeronautics and Astronautics Space 2011 Conference*. AIAA 2011-7185, (pp. 1–24). http://www.spudislunarresources.com/Bibliography/p/102.pdf.

Sterns, P. M., Tennen, L. (2002). Privateering and profiteering on the Moon and other celestial bodies: Debunking the myth of property rights in space. In *Proceedings of the forty-fifth colloquium on the law of outer space* (pp. 56–65).

Stovitz, C. J., Loomis, T. (1985). Space law: Lessons learnt from the Antarctic, *Proceedings 28th Colloquium* (pp. 165–171). Stockholm.

Sunday Times (2013). *China and India in Race to Exploit the Indian Ocean Seabed.* http://www.sundaytimes.lk/131215/sunday-times-2/china-india-in-race-to-exploit-indian-ocean-seabed-76395.html. Retrieved 24 January 2015.

Taylor, G. J. (1985). The need for a lunar base: Answering basic questions about planetary science. In W. W. Mendell (Ed.), *Lunar bases and space activities of the 21st century* (pp. 189–197). Houston: Lunar and Planetary Institute.

Taylor, S. R. (1998). *Destiny or Chance? Our solar system and its place in the cosmos*. Cambridge: Cambridge University Press.

Toon, O. B. (1997). Environments of earth and other worlds. In: *Carl Sagan's Universe.* Cambridge: Cambridge University Press.

Tronchetti, F. (2009). *The exploitation of natural resources of the Moon and other celestial bodies: A proposal for a legal regime*. Leiden: Martinus Nijhoff Publishers.

Tronchetti, F. (2014). Private property rights on asteroid resources: Assessing the legality of the ASTEROIDS act. *Space Policy, 30*, 193–196.

UN (1984). The United Nations treaties on outer space. New York: United Nations.

UN (1999a). *Report of the third united nations conference on the exploration and peaceful uses of outer space. Vienna*, 19–30 July 1999. Cooperation Among Nations point 5, p. 110. http://www.oosa.unvienna.org/pdf/reports/unispace/ACONF184_6E.pdf.

UN (1999b). Technical report of the space generation forum (pp. 14–15). http://www.caisu.org/nsaw/2004/img/sgf_technical_report_july_1999.pdf.

UN Office of Outer Space Affairs. (1999). *International cooperation in space activities*. http://www.unoosa.org/oosa/en/unisp-3/index.html.

Unispace III (1999). Participants of the space generation forum. http://www.un.org/events/unispace3/pressrel/sgforum/sgfattend.htm.

United Nations Division for Ocean Affairs and the Law of the Sea, Office of Legal Affairs, and the International Seabed Authority. (2013). *Marine Mineral Resources Scientific Advances and Economic Perspectives.* http://www.isa.org.jm/files/documents/EN/Pubs/ISA-Daolos.pdf.

Van Dyke, J. M. (1988). *International navigation: Rocks and shoals ahead?* Honolulu: Law of the Sea Institute, University of Hawaii.

Viikari, L. (2012). Natural resources of the Moon and legal regulation. In V. Badescu (Ed.), *Moon: Prospective energy and material resources* (pp. 519–552). Heidelberg: Springer.

Wolfrum, R. (2010). Legitimacy of international law and the exercise of administrative functions: The example of the international seabed authority, the International Maritime Organization (IMO) and international fisheries organizations. In A. von Borgondy, et al. (Eds.), *The exercise of public authority by international institutions*. Berlin: Springer.

Zhukov, G., & Kolosov, Y. (1984). *International space law.* New York: Praeger Publishers.

Chapter 13
Rendezvous with Research: Government Support of Science in a Space Society

James S.J. Schwartz

Abstract Mark Brown and David Guston have outlined a framework for addressing state protection of scientific freedom in the contemporary United States, arguing that a limited freedom of scientific research improves democratic governance. I show how this reasoning, with modification, can be applied to protection of scientific freedom in a democratic lunar society, and I argue that lunar society will experience a comparatively greater need to ensure scientific freedom.

Keywords Space science · Science governance · Democracy scientific freedom

13.1 Surface Detail: Introduction

In this essay I discuss state support of scientific research in a democratic space society, e.g., a democratic lunar society. I contend that much of what we today describe as "space" science, in a space society, will make vital contributions to democratic governance. Therefore a democratic space society will have compelling reasons for supporting a great deal of scientific research (though the specifics inevitably will vary from society to society). This pattern of reasoning is not unique to the issue of state support of science in a space society—indeed, I am self-consciously borrowing from recent discussions of state support of science in the contemporary United States. Nonetheless, I argue that the scope and force of this

Thanks to the following individuals for discussion and comments: Charles Cockell, Tony Milligan, John Rummel, Annalea Beattie, Erik Persson, Janet De Vigne, John Carter McKnight, Euan Monoghan, and Mukesh Bhatt. Thanks also to Robert Feleppa for commenting on an earlier draft.

J.S.J. Schwartz (✉)
Department of Philosophy, Wichita State University, Wichita, USA
e-mail: james.schwartz@wichita.edu

C.S. Cockell (ed.), *Human Governance Beyond Earth*,
Space and Society, DOI 10.1007/978-3-319-18063-2_13

kind of reasoning is greater in the case of a democratic space society than in the case of the United States. After all, so much more is known about the Earth, its resources, and its ability to support human life, than is known about any other location in the Solar System (and its resources and ability to support human life).

Recently the issue of state support of science has been addressed as a question about constitutional protection for scientific freedom, where scientific freedom is taken as a *negative* liberty that one has when the state imposes minimal restrictions on permissible research projects. This negative scientific liberty, which falls under the more general heading of free inquiry, is often taken for granted as both a precondition and an adjuvant for democratic deliberation. But, as I discuss in Sect. 13.2, recent science policy work suggests that inquiry (broadly construed) is subject to decidedly weak protection, with the strongest protection available only for lines of inquiry which directly impact democratic deliberation. I will often refer to this position as the Brown-Guston position or framework [which is drawn from (Brown and Guston 2009)]. In order to address the *positive* liberty of government support or subsidization of science, I amend the Brown-Guston position by assuming that democratic societies are *obligated* to support activities which are likely to make important contributions to democratic governance. Thus, I claim, the Brown-Guston position provides a framework for addressing not only scientific freedom of *ends* but also scientific freedom of *means*.

In Sect. 13.3 I argue that the Brown-Guston position (along with my amendment) provides comparatively greater protection of (and support for) science in a lunar society. Most of what we would call "lunar science" would, on the Moon, be regarded as ordinary but vital research for the maintenance of lunar society. After all, an extensive understanding of the physical conditions of one's environment is an adjuvant (if not also a precondition) for democratic deliberation. Given this broader support for research, the Brown-Guston position is worth considering as a model for the state support of science in a lunar society. This raises the question as to which lines of inquiry are most worthy of the support of the lunar state. In response I engage in moderate speculation concerning the extent to which current research proposals map onto the anticipated needs and ambitions of lunar society.

I conclude the essay in Sect. 13.4 by discussing an objection to my use of the Brown-Guston position, which was developed with the intention of addressing *restrictions* on scientific freedom of ends, e.g., whether the state can cave into public pressure and ban so-called "controversial" research projects such as human embryonic stem cell research. Space science, fortunately, tends not to offend moral sensibilities, and is therefore unlikely to require the kinds of constitutional protections sometimes necessary for "controversial" research. Moreover, many democratic societies support science, including space science, well beyond the requirements of the Brown-Guston position. Nevertheless I maintain that the Brown-Guston position, regardless of its intended application, remains relevant to the discussion at-hand because it provides a framework for circumscribing the minimum acceptable degree of support for science in a democratic lunar society.

13.2 A Gift from Earth: Democracy and Scientific Freedom

In this section I explore the political grounds for government support of science in a democratic society—where by 'support' I mean the (full or partial) provision of resources necessary for conducting scientific research—which I will sometimes refer to as the "freedom of means" of science. In the language of liberty, then, I am interested in the grounds for including scientific research in the *positive* liberties guaranteed in a space society. However, at least in the philosophical and science policy literature, this issue attracts comparatively little discussion. More prominent is discussion concerning the grounds for government protection—in particular, *constitutional* protection—of scientific *freedom*, where 'freedom' consists in being able to determine, without state interference, the *ends* of one's research (which I will sometimes refer to as the "freedom of ends" of science). In the language of liberty, then, extant discussion aims to circumscribe the *negative* liberties associated with scientific research. It should be clear that the absence of prohibitions on undertaking a course of action—scientific research included—does not necessarily imply a concomitant entitlement to the means to carry out that course of action. I might be free to study philosophy of mathematics, but that does not mean I am entitled to no-cost access to the latest books in the field. Similarly, a researcher might be free to investigate the social habits of readers of Iain Banks novels, but that does not mean she is entitled to the resources necessary for conducting this research.

Nevertheless determining the scope of governmental/constitutional protection of scientific freedom of ends can help to illuminate the grounds for government *support* of science (scientific freedom of means). If a democratic society is at all obligated to arrange for governmental (or perhaps constitutional) *support* for science, then those research projects thusly supported should comprise a subset of the research projects which citizens are constitutionally *free* to pursue.[1] According to recent work,[2] which I discuss below, protection for scientific freedom of ends extends primarily to research projects which are likely to contribute to democratic deliberation (e.g., through enabling or informing the democratic process). This point permits a more ambitious conclusion when coupled with the additional assumption that democratic societies are obligated to carry out those activities which are required for enabling or informing the democratic process: Namely, that a democratic society is obligated to ensure freedom of ends *and* means for those projects which are likely to contribute to democratic deliberation.[3] What this suggests, *prima facie*, is that the projects protected modulo scientific freedom of ends

[1]One kind of exception here is weapons research (there are likely to be other exceptions). Governments might have compelling interests to (a) support weapons research, but also to (b) prohibit citizens from freely engaging in such research.

[2]Brown and Guston (2009), Weinstein (2009), and Wilholt (2010).

[3]Ignoring for the duration the "ought implies can" objection that arises whenever there are too many promising projects and too little available funding.

coincide with the projects that a democratic society is obligated to support (i.e., protected under scientific freedom of means). According to this picture, then, scientific freedom (of means *and* ends) should be ensured at least to the extent that these liberties are necessary for the smooth-functioning of society.[4]

Before beginning in earnest, let me note that this discussion is self-consciously directed at a constitutional democracy not unlike that of the United States. In particular, the dialectical strategy under consideration subsumes scientific research under the free speech clause of the First Amendment to the United States Constitution,[5] which in effect judges research activities differently depending on whether they are classified as speech, expressive conduct, etc. Consequently, United States constitutional law decisions comprise much of the evidence used by the authors under consideration. This strategy limits the generalizability of this section—no two democratic societies are perfectly alike. However, as I shall argue in Sect. 13.3, this strategy does promise to provide fairly strong support for science in a democratic *space* society, regardless of what kind of support it entails for science in the present-day United States. It is therefore worth considering as a model for the support of science in a space society.

13.2.1 Democracy and Scientific Freedom of Ends

At first blush one might suppose that the democratic case for scientific freedom of ends is a simple matter of categorizing science as a kind of free speech or expression, or as a variety of conduct essential for free speech or expression. Scientific research, then, would presumably inherit the same constitutional protections that are provided for speech. However, constitutional protection for speech comes in degrees, and according to recent analyses, the strongest protections are given to speech (or conduct) that either enables, informs, or is constitutive of the democratic process.[6] According to this reasoning, rights—including the right to free speech—acquire their force not as "natural, pre-political, individual protections against society and the state," but rather, "as a result of concrete struggles for social equality" (Brown and Guston 2009, pp. 356–357). In other words, rights function to promote social equality, and in turn, to improve democratic governance. Thus the First Amendment to the United States Constitution protects free speech principally because such a freedom is (at least in principle) a means for promoting equality and

[4]I will not discuss the implications of this reasoning for *private sector* science, but I should not in any way be taken as suggesting that scientific freedom in the private sector is of little consequence. For discussion see Frankel (2009).

[5]"Congress shall make no law respecting an establishment of religion, or prohibiting the free exercise thereof; or abridging the freedom of speech, or of the press; or the right of the people peaceably to assemble, and to petition the government for a redress of grievances."

[6]Below I will emphasize the position of Brown and Guston (2009), but *cf.* Weinstein (2009) and Wilholt (2010).

proper democratic governance. Therefore speech acts (and forms of expression) are subject to protection in proportion to their service (actual or potential) to democratic decision-making. In contrast, speech (and expression) that lies far from the core of democratic decision-making can be subject to extensive regulation.

What does this mean for scientific freedom of ends? The first thing to note is that, in treating scientific research as a kind of free speech, one loses any special protection for scientific speech, expression, or conduct. What is protected, if anything, in a more general right to free inquiry (and associated speech, expression, and conduct) (ibid., p. 359). This is because there is no constitutionally recognized category of scientific speech or research (Post 2009, p. 431), which reflects the egalitarian nature of the United States Constitution. Scientists are not privileged members of society—a right specifically to *scientific* inquiry would unduly benefit those with the requisite interests and resources for conducting scientific research.

The second thing to note is that, just as speech and inquiry are protected in proportion to their contributions to democratic decision-making, so consequently is scientific freedom of ends. This raises the unenviable task of determining which forms of inquiry are essential or are likely to contribute to democratic decision-making. According to Brown and Guston, this includes inquiries that will, "help resolve political questions," "contribute to [democracy's] material preconditions (e.g., economic growth, scientific literacy, health, physical security)," and "support democratic culture, broadly conceived, simply by satisfying intellectual curiosity or offering aesthetic pleasure." (2009, pp. 362–363). Helping resolve political questions is most integral to the democratic process, they insist, and so inquiries which help resolve these questions are subject to the strongest protection. The other two categories of inquiry are subject to "commensurably less protection" presumably because they are less integral to the democratic process (ibid.).[7]

It is difficult to state in the abstract precisely what forms of inquiry, and what forms of scientific inquiry in particular, are protected under this position. This point is not offered as a criticism of the Brown-Guston position, which does not in itself aim to settle questions about the strength of protection available to particular lines of inquiry—but it raises the question nonetheless.[8] There are some readily discernible answers—certainly the studies of climatology, medicine, and alternative energy are strongly protected, as each line of inquiry promises to help answer political questions and to contribute to the material preconditions of democracy.

But these answers, though certainly not exhaustive, are unhelpfully general. It would not be plausible to maintain that *all* climatology studies are protected, just as it wouldn't be plausible to insist that all medical and alternative energy studies are

[7]I will not judge whether this is true of the contemporary United States. But I will point out that this ordering is contingent—it is possible that there exist societies for which it is most important for research to satisfy intellectual curiosity, for example. In Sect. 13.3 I will revisit the issue of ordering for lunar society.

[8]Brown and Guston do suggest that the social sciences are more likely to impact democratic deliberation, which "leads to the perhaps surprising conclusion that in some cases the social sciences may have a stronger claim than the natural sciences to a right to research" (ibid., p. 363).

protected. It simply isn't clear what granularity best suits this discussion. There are difficulties all around. Judging the merits of scientific disciplines is complicated by (a) the difficulty of reliably and naturally demarcating disciplines (difficulties which get compounded as one goes from discipline to sub-discipline, and from sub-discipline to sub-sub-discipline, etc.), and by (b) the fact that this level of analysis is far too general to distinguish between worthwhile and fruitless research projects. On the other extreme, judging the merits of individual experiments is (c) hopelessly cumbersome, and (d) prone to ignoring or trivializing the context in which the experiment is to take place. An apparent compromise is to adjudicate merit at the level of what Kitcher (2001) would describe as *significant questions*—questions which reflect the interests of society and which interlink with science in various ways.[9] But questions—significant or mundane—are likely to link up with a very wide array of scientific disciplines and research projects. A simple interest in a particular kind of bacterial infection might implicate inquiries spanning from genetics to molecular biology to chemistry to optics to physics. Scientific research is full of serendipitous interconnections and dependencies—disciplines, subdisciplines, and research projects generally cannot be pursued in isolation.

One could attempt to avoid the difficulties just mentioned by insisting on a broad protection for inquiry—including scientific inquiry—by advancing the following simple-minded argument[10]:

> Freedom of inquiry—but freedom of scientific inquiry most especially—is the optimal setting from which to grow our collective knowledge. Growing our collective knowledge is essential for democratic decision-making. Thus, freedom of scientific inquiry is the optimal setting for a democratic society.

But this argument is open to equally simple-minded objections, perhaps the most salient being that not all knowledge acquisitions are relevant to democratic decision-making. One way to grow our collective knowledge is for all of us to memorize our own clothing-permutations given our current wardrobes. But this activity would be clearly inessential to the democratic process (one hopes!), not to mention a colossal waste of time and energy. So it's simply wrong to say that inquiry broadly speaking is essential to democracy—some inquiries will be, others will not.

We also should resist the temptation to appeal to democratic deliberation itself to resolve this difficulty. After all, democratic deliberation is all too easily corrupted by misinformation, ignorance, motivated reasoning, etc. As I see it, what underlies the Brown-Guston framework is that scientific *freedom* makes vital contributions to democratic governance. It is essential here that the science be *free*—and it cannot be free if it is subject to the whims of the majority. It is true that this is a restricted

---

[9]See, e.g., Kitcher's "significance graph" for the question "how do organisms develop?" (ibid., p. 79).

[10]This argument is a variation on what Wilholt (2010) and Bayertz (2006) have called the "epistemic argument" for scientific freedom.

freedom—it does not extend to every conceivable research project, and it is bounded by fiscal and ethical constraints. But that does not prevent us from treating it as an objective matter whether and to what extent research projects help resolve political problems and contribute to the material preconditions of democracy. If there is disagreement about the proposed utility of a line of research, then let the evidence be presented and evaluated by appropriate authorities (e.g. scientists, philosophers of science, and science policy theorists). It remains an open (though time-indexed) task to determine the merits of many research projects—one that I will not attempt to complete.

Although we cannot easily deduce every project worthy of pursuit, neither are we entirely in the dark. We do know something about the needs and problems of society. We also know something about how to meet those needs and answer those problems. And we also know something about how science can play a role in the solution or dissolution of those problems. We can in turn use this information to produce more or less informed speculation regarding what forms of inquiry are most likely to contribute to democratic deliberation, at least to the extent we are aware of the research possibilities. This is precisely what I propose to do in later sections of this essay—to engage in informed speculation regarding space science's contributions to democratic deliberation. Nevertheless I do not propose to come to any decisive answers concerning the moral or political status of the space sciences —the ensuing discussion represents the *beginning* of a conversation rather than its conclusion. Which is all to say that while I acknowledge the difficulties associated with subjecting the merits of inquiry to democratic values, that should not prevent us from examining how these values play out when applied to space science. However, heretofore this discussion has been restricted to scientific freedom of *ends* —but it is possible to connect this negative liberty to the positive liberty of scientific freedom of *means*.

13.2.2 From Scientific Freedom of Ends to Scientific Freedom of Means

In order to stake out a position according to which a democratic society is obliged to ensure at least some degree of freedom of scientific *means*, I must extend the position of Brown and Guston. I will do this by assuming that a democratic society is obligated to support those activities that are likely to contribute to democratic deliberation. I hope this assumption strikes the reader as intuitively plausible, and in any case I will not argue for it. This obligation, like most obligations, is not absolute; the state has duties other than those associated with facilitating democratic deliberation, duties which may in some cases override its duty to facilitate democratic deliberation. With this assumption in hand we can offer the following argument for scientific freedom of means:

1. A democratic society is obliged to do *X* if *X* is likely to contribute to democratic deliberation.
2. Ensuring freedom (of ends and means) for at least some forms of scientific research is likely to contribute to democratic deliberation.[11]
3. Therefore, a democratic society is obliged to ensure freedom (of ends and means) for at least some forms of scientific research.

We should first observe that the strength of the obligation given in the conclusion is tied to the degree to which scientific research contributes to democratic deliberation. Thus, according to the Brown-Guston position, the strongest obligations are tied to inquiries which help resolve political questions, with weaker obligations tied to inquiries which merely contribute to the material preconditions of democracy or to democratic culture.[12]

We are now in a position to comment on what is implied regarding government support of *space* science. Whether there should be freedom of ends—and hence, freedom of means—for space science turns on the degree to which space science helps resolve political questions, contributes to the material preconditions of democracy, and supports democratic culture by satisfying intellectual curiosity or offering aesthetic pleasure. There are certainly forms of space science that fall under each category:

Resolving political questions: Earth observation from space helps resolve political questions by providing an extraordinary amount of policy-relevant information about weather, climate, resources, development, etc. Moreover, some forms of planetary science, e.g. comparative atmospheric studies, might prove invaluable in our attempts to mitigate anthropogenic climate change. But most other forms of space science are largely irrelevant to political questions (except when the question is whether to fund space science, which trivially applies to all forms of space science).

Contributing to the material preconditions of democracy: Recall that scientific research contributes to the material preconditions of democracy when it facilitates economic growth, scientific literacy, human health, and national security—and one of the most common justifications for space expenditures is that space science will contribute to these areas. Space-based laboratories offer unique opportunities for life and materials sciences, which could potentially lead to the manufacturing of medicines and materials in space. Scientific research will also be necessary in advance of the exploitation of, e.g., lunar and asteroid resources. Further disciplines implicated here include astrobiology, astronomy, chemistry, geology, planetary science, and physics. However, it is uncertain what contributions these

[11]This premise is stated more weakly than is strictly necessary, because we know that many forms of scientific research are not just likely to but do in fact contribute to democratic deliberation.

[12]One worry here concerns the breadth of support necessary for genuine scientific freedom of means for some particular area of research. Would a government satisfy its obligation to ensure this freedom by supporting a single individual's research? Or, in contrast, could a government only satisfy its obligation to ensure this freedom when all interested parties are provided the means to conduct research in the area? Neither extreme is acceptable. What will be required for this position to work is (a) some way of determining *how much* research activity is reasonable for a particular kind of inquiry, and (b) a way of ensuring that, once the means are provided, the remaining scientific work is free from state control (beyond uncontroversial ethical demands, e.g. not to abuse research subjects).

sciences will make to economic growth, human well-being, etc. Let me raise one brief concern.[13] If the exploitation of space resources is to be conducted as an extension of current market activities, it is quite likely that the lion's share of any resulting economic growth will benefit the already wealthy, and thus will not lead to much improvement in the well-being of average citizens. Therefore we have some reason for doubting that exploitation-driven space science will contribute to economic growth in ways that promote—rather than diminish—social equality, and in turn, the democratic process.[14]

Satisfying intellectual curiosity and offering aesthetic pleasure: Virtually all forms of space science could be described as satisfying intellectual curiosity. As well, there are forms of space science—most notably astronomy and Earth observation—that offer aesthetic pleasure by providing sublime images of our planet, solar system, and universe. The comparative value of such contributions is difficult to ascertain, but this value is non-negligible, as evidenced by the widespread public opposition to recent proposals to shut down the Hubble Space Telescope.[15]

Recall that, as we go down the list, the strength of the obligation to ensure freedom of ends (and means) decreases. Thus, although virtually all forms of space science satisfy intellectual curiosity, this does not count very strongly in their favor (except in situations of extreme prosperity). Given the large costs of operating in space, surely we could satisfy more intellectual curiosity by funding greater quantities of cheaper scientific inquiries. And, as I have suggested above, I am not entirely confident that space science's contributions to the material preconditions of democracy will count especially strongly in its favor. Which is all to say that under my extension of the Brown-Guston framework, the most compelling support for space science (at least given the political realities of the present-day United States) is available to those research projects which are likely to help answer political questions.

The implication here is that scientific freedom of means (and ends) for space science is quite limited, extending not far beyond those research projects which help resolve political questions, most notably Earth observation projects.[16] However, as the reader has surely noticed, although this result may be valid for the contemporary United States, results will be entirely different when we apply the Brown-Guston framework to a democratic society in space, e.g., a democratic lunar society. In the next section I shall offer reasons for supposing that a democratic lunar society will

[13]Which I explore in more detail in (Schwartz 2014).

[14]For discussion about the impact of space science on scientific literacy in the United States, see Fraknoi (2007).

[15]All of these claims are complicated by the fact that scientific research is serendipitous—and it is in general not possible to predict what research will produce answers to what questions. Nor is it possible to predict what research will lead to the development of socially and economically fruitful methods and technologies. And nor is it possible to predict what research will become infused into lay culture. We might well decide to take this as sufficient justification for wide-ranging support of pure, autonomous scientific research—a position that is well-represented in the work of Gonzalo Munévar.

[16]For a related judgment, see Wilholt (2006).

require a more expansive protection for freedom of scientific means (and ends), largely because a lunar society will experience comparatively greater needs with respect to ensuring the material preconditions of democracy and with respect to promoting democratic culture.

13.3 Revelation Space: The Democratic Case for Science in a Space Society

If the reasoning of the previous section is to be trusted, the relevance of space science to democratic deliberation, and hence the ground for its protection by the state, diminishes quite rapidly as we consider research activities taking place further and further from Earth. But things will look very different when we contemplate the deliberative needs of a space society, e.g., an independent lunar society. Indeed, much of what we today describe as "space science" will simply be called "science" by denizens of a space society—and much of this science will prove essential for answering political questions and for ensuring the material preconditions of democratic deliberation. In other words, we can expect the Brown-Guston justification for limited scientific freedom of ends (and by my extension, scientific freedom of means) to be much less limited in the context of a democratic space society.

How does the Brown-Guston reasoning play out in space settlements? Given the diverse possibilities of location and of governance that space settlements provide, it is not possible to answer this question in full generality, so I will restrict my remarks to a lunar society which (quite conveniently) has adopted a constitutional democracy similar to that of the United States. But I believe that the idiosyncrasies in what follows are primarily tied to the location of the society—the Moon—rather than the form of governance the society adopts. Space societies, regardless of their form of governance, will experience great challenges in providing for the material needs of their citizens—shelter, water, food, air, etc.—needs which can be so readily met on Earth, but not in space. Therefore all space societies, regardless of their form of governance, will share a compelling interest in supporting the kinds of research necessary for developing and improving the technologies required for securing the basic provisions of life. But the particulars will vary based on the location of the society—the knowledge and technical capabilities needed to survive on the Moon will be different from the knowledge and technical capabilities needed to survive on Mars or on Ganymede (which is not to deny that in learning how to survive on the Moon we will come to know more about how to survive on Mars and on Ganymede).

We can now ask a more refined question—how is the Brown-Guston reasoning likely to play out in the case of a democratic lunar society? What should appear obvious by now is that Brown and Guston's prioritizing of answering political questions over ensuring the material preconditions of democratic deliberation seems inappropriate for a lunar society (at least in its early stages of development). Since the most pressing issues lunar society will face concern basic life-giving resources, it will be most incumbent on lunar society to ensure the material preconditions of democratic deliberation—political questions will be either secondary or derivative. That is not to discount political questions, including those not directly related to ensuring survival[17]—just to note that they are not of unparalleled importance.

There is moreover reason for placing a greater emphasis on providing opportunities to satisfy intellectual curiosity and to experience aesthetic pleasure. This is because the lunar environment does not naturally provide such opportunities to anything approaching the degree they are available on Earth. Consider living in a lunar settlement. In contrast to life on Earth, life would probably seem extremely dull and confining. Even if the average citizen had the ability to venture beyond the confines of the settlement, they would be confronted by the rather bleak and homogeneous lunar surface.[18] Charles Cockell argues that to escape from this "extraordinary banality in the day-to-day experiences of weather, climate, and general environmental changes," we should encourage "intellectual efflorescence" (2008, p. 270). I would make an even stronger claim: without such opportunities for intellectual efflorescence, life in a lunar colony would simply not be worth living. That scientific freedom can create such opportunities counts strongly in its favor—and I would insist that the humanities are implicated here in addition to the natural sciences.

The upshot is that lunar society will first and foremost be compelled to satisfy the material preconditions of democracy; placing somewhat less emphasis on resolving political questions and contributing to democratic culture. Nonetheless lunar society's obligation to contribute to its cultural development is comparatively greater than, e.g., the US government's obligation to contribute to its society's cultural development. Thus the potential for a scientific discipline to make cultural contributions to lunar society is not of little moment.

What therefore can we say about the extent of scientific freedom in lunar society? What research is likely to help ensure the material preconditions of democracy, resolve political questions, or contribute to democratic culture? To help answer these questions I shall review the research projects identified in Crawford et al. (2012), which, according to its title, aims to provide a "scientific rationale for resuming lunar surface exploration." (I remind the reader—this is the beginning of a discussion, and not its end. As we come to understand more about the Moon and its ability to support human settlement, we will no doubt improve our understanding of what research is most likely to facilitate democratic deliberation and to fulfill other societal goals.)

---

[17]After all, we can also be confident that lunar citizens will experience challenges to personal liberty that might seem unacceptable to those accustomed to modern occidental living—especially assuming that lunar citizens will reside in cramped subsurface habitats. See Cockell (2013) for an extensive discussion of questions related to personal liberties in space.

[18]Though perhaps it only seems bleak and homogeneous to those of us not reared on the Moon.

Early solar system evolution and planetary-geological evolution: Since it is virtually geologically static, the Moon can be studied as a repository of the early history of the solar system. Materials on and below its surface can date as far back as the creation of the Earth-Moon system (some 4.5 billion years ago). Therefore, studying the Lunar surface and subsurface will vastly extend our knowledge of the bombardment history of the inner solar system (ibid., p. 4), as well as our understanding of the habitability of the Earth since its origin (ibid.).

At first blush these projects appear to be of strictly academic interest. Does the study of early solar system evolution contribute to democratic deliberation in further ways? It can.[19] A Lunar society will almost certainly depend on in situ resource utilization, which will require, at a minimum, understanding what resources are distributed in what fashion. The geological study of the Moon—whether conducted as an exercise in understanding the history of the solar system or otherwise—will provide information about the physical distribution of Lunar resources. This information will be important not only for satisfying the material needs (and wants) of lunar citizens, but also for resolving political questions about resource use and conservation. Thus there should be freedom of ends and means for research on early solar system evolution.

Mineralogy: A proper understanding of the distribution and diversity of lunar crystal rocks, regolith, and other minerals will require extensive surveying and analysis— especially on the relatively unexplored lunar farside and poles (ibid., p. 5). This information is clearly of significance for in situ resource utilization, and as such, will, just as early solar system and planetary-geological evolution, aid in the satisfaction of the needs and wants of lunar citizens, as well as help resolve political questions about resource use and conservation.

Volatiles: Observations from lunar orbit indicate the presence of water ice in certain craters of the lunar poles, but the actual extent of the deposits is unknown (ibid., pp. 5–6). Given that human existence and flourishing depend on access to sufficient quantities of water, an understanding of the origin and extent of lunar water ice deposits would be crucially important for a lunar society—at least assuming that such a society cannot easily access external sources of water. As with mineralogy and early solar system evolution, the study of lunar volatiles will help satisfy the material needs of lunar citizens and will help answer political questions.

Human physiology and medicine: We understand quite little about the ability of human beings to survive, reproduce, and thrive in zero-and reduced-gravity environments (ibid., p. 8). It will be incumbent upon a lunar society to develop a sufficient understanding of the effects of lunar gravity on citizens in order to ensure the long-term health and sustainability of its human population (as well as the health and sustainability of any other Earth-organisms on which the human population depends). Thus, studying human physiology and medicine rates very highly for its contributions to the material preconditions of lunar democracy.[20]

Astrobiology: Examining the lunar geological record for evidence of organic material will improve our understanding of how life might have developed on Earth. As well, studying the effects of the lunar environment on microbial life (including studying the crashed remains of unsterilized spacecraft) will improve our understanding of the conditions under which we may discover life elsewhere (ibid., pp. 7–8). This research will help lunar citizens come to understand how the lunar environment must be modified or supplemented in order to support life, and is therefore relevant to developing, maintaining, and improving

[19]Thanks to John Rummel for the following observation.

[20]A related ethical question that I do not have space to discuss is the following: Would it be permissible in the first place to establish a lunar society *without* knowing beforehand that human beings are capable of living and reproducing in lunar gravity? Perhaps this is the kind of research that ought to be conducted *before* the establishment of a lunar society.

agricultural and other life support systems on the Moon. Thus, astrobiology promises to contribute to the material preconditions of lunar society and to help answer political questions.[21]

Astronomy: The Moon, with its negligible atmosphere, is an exceptional location for making a wide variety of astronomical observations—especially the lunar farside, which is naturally shielded from Earth's radio emissions. Astronomical observations from the Moon could, depending on the devices employed, improve our understanding of the evolution of the universe, expedite the search for radio emissions from exoplanets, and detect cosmic ray particles (ibid., 10). It is however difficult to see how lunar astronomy would help answer political questions or contribute to the material preconditions of democracy; rather, lunar astronomy contributes mostly to the satisfaction of intellectual curiosity.[22]

We can see then that Brown-Guston style reasoning affords fairly broad protection and support for science in a lunar society.[23] Moreover, given the increased importance of providing opportunities for experiencing aesthetic pleasure and satisfying curiosity, lunar society will have a non-negligible obligation to support research in fields such as astronomy, which are unlikely to make direct contributions either to the material preconditions of democracy or to the resolution of political questions.

I think we have arrived at a fairly pervasive grounding for scientific freedom (of ends and means) in a lunar democratic society. However, we should not forget that we are contemplating the establishment of an entirely new society, and one that is rooted in an historically novel location—the Moon. Although political precedents from Earth may yield insights into what forms of governance successfully uphold democratic ideals, nonetheless these precedents need not categorically determine rights and responsibilities within lunar society. As we have already seen, lunar society will need to place comparatively greater emphasis on contributing to democratic culture. Furthermore, as Cockell has demonstrated in detail, the lunar environment is especially conducive to tyrannical control over basic life-giving resources, given that oxygen and water must be manufactured (Cockell 2008, 2009). A specific constitutional right to oxygen verges on nonsense for present-day Earth, but it might be absolutely necessary for ensuring minimal personal liberties on the Moon.[24] The simple point here is that what is necessary for smooth-functioning democratic governance in a lunar society will be quite different from what is necessary for smooth-functioning democratic governance on Earth, whether we are talking about basic rights, or about science.

---

[21]Thanks to Charles Cockell for discussion about the political relevance of astrobiology.

[22]On the other hand, if lunar society comes to depend for its existence on the exploitation of asteroid material, then it would have further and stronger reasons for ensuring freedom of astronomical research.

[23]To this list I would add economics, social science, and psychology, as these fields also appear to be essential for devising the social and political tools necessary for the maintenance of a lunar society.

[24]Of course, some reasoning derived from the Brown-Guston framework can easily accommodate the case of oxygen. Access to oxygen is a material precondition of democratic deliberation, and thus, citizens should have an extremely strongly protected right to such access.

We should also be prepared for a more radical departure from the assumptions underlying the Brown-Guston framework. Not only should we expect that the precise requirements for promoting democratic ideals will vary according to the environmental demands lunar society faces—we should also acknowledge that lunar society may wish to codify its own set of ideals—ideals different from those exemplified in the US constitution. For instance, lunar society may decide to place direct value of scientific research, even if this means supporting research that is incredibly unlikely to contribute to democratic deliberation. This would be a prudent move, in that it would provide another ingress point for infusing science into the culture of lunar society. After all, lunar society will need as many buffers as possible against scientific illiteracy, given the importance of scientific knowledge to survival, flourishing, and freedom in space (Cockell 2009, p. 146).

13.4 Divergence: Conclusion

The Brown-Guston framework provides fairly substantial support for science in a space society. Therefore it is worth considering as a model for thinking about government subsidization of scientific research in a democratic lunar society. To conclude this essay, I would like to address a feature of the Brown-Guston framework which might be thought to unveil it as an inappropriate model for thinking about scientific freedom of *means*.

As I have mentioned previously, the Brown-Guston framework explicitly addresses scientific freedom of *ends*, and is specially tailored to address questions about the legality of allegedly "controversial" lines of research, such as research involving human embryonic stem cells. One could describe Brown and Guston's project as an attempt to circumscribe the areas of research which should be protected (due to their importance to society) even if they are objectionable in the eyes of the public. But the bulk of important scientific research, including space science, does not raise these kinds of concerns (except from fringe anti-science/anti-technology interest groups).[25] For this reason one might question whether the Brown-Guston framework is appropriate when applied to research projects which are not likely to test the moral patience of the wider public.

However, I do not believe the above concern indicates a defect in applying the Brown-Guston framework to science in a lunar society. First, we should note that, in point of fact, governments provide much greater support to science than is demanded under the Brown-Guston framework. We do not support only research that is politically expedient but a great deal of further research, pure and applied.

[25]Some of the more common objections to the state support of science, including state support of space science, include the humanitarian objection that (a) money spent on science could be better spent elsewhere, and the libertarian objection that (b) the state is not an appropriate agency to forward the progress of science. The libertarian objection is nowadays often raised against government-funded space science. Here is not the place to counter these objections.

Rather, what the Brown-Guston framework helps us identify is the minimum acceptable commitment to science in a democratic society. It is reasonable to suppose that lunar society, once it reaches a certain level of economic and social stability, will provide a more expansive support for science (as well as the arts and humanities) than is demanded under the Brown-Guston framework. Moreover, although most (if not all) of space science avoids offending moral sensibilities, we cannot guarantee that it will continue to avoid doing so in the future. Perhaps, despite our best efforts, there will be a time when Young Moon Creationism takes hold in lunar society, with a large percentage of the public calling for the abandonment of most science activities. The Brown-Guston framework provides a constitutional basis for combating this—pardon the non-pun—*lunatic* influence.

References

Bayertz, K. (2006). Three arguments for scientific freedom. *Ethical Theory and Moral Practice, 9*, 377–398.

Brown, M., & Guston, D. (2009). Science, democracy, and the right to research. *Science and Engineering Ethics, 15*, 351–366.

Fraknoi, A. (2007). Space science education in the United States: The good, the bad, and the ugly. In Dick, S., & Launius, R. (Eds.), *Societal impact of spaceflight* (pp. 407–419). NASA Publication SP-2007-4801.

Cockell, C. (2008). An essay on extraterrestrial liberty. *Journal of the British Interplanetary Society, 61*, 255–275.

Cockell, C. (2009). Liberty and the limits to the extraterrestrial state. *Journal of the British Interplanetary Society, 62*, 139–157.

Cockell, C. (2013). *Extraterrestrial liberty: An Enquiry into the nature and causes of tyrannical government beyond the earth*. Edinburgh: Shoving Leopard.

Crawford, I. A., et al. (2012). Back to the moon: The scientific rationale for resuming lunar surface exploration. *Planetary and Space Science, 74*, 3–14.

Frankel, M. (2009). Private interests count too. *Science and Engineering Ethics, 15*, 367–373.

Kitcher, P. (2001). *Science, truth, and democracy*. New York: Oxford University Press.

Post, R. (2009). Constitutional restraints on the regulation of scientific speech and scientific research. *Science and Engineering Ethics, 15*, 431–438.

Schwartz, J. (2014). Prioritizing scientific exploration: A comparison of the ethical justifications for space development and for space science. *Space Policy, 30*, 202–208.

Weinstein, J. (2009). Democracy, individual rights and the regulation of science. *Science and Engineering Ethics, 15*, 407–429.

Wilholt, T. (2006). Scientific autonomy and planned research: The case of space science. *Poiesis & Praxis, 4*, 253–265.

Wilholt, T. (2010). Scientific freedom: Its grounds and their limitations. *Studies in History and Philosophy of Science Part A, 41*, 174–181.

Chapter 14
Art Making, Good Governance and Sustainable Extra-Terrestrial Communities

Annalea Beattie

Abstract In conditions of extreme adversity in space, resilient communities and robust organisations will only develop and perpetuate if people live communally. Of specific relevance to hardy extra-terrestrial societies who value collective long-term growth is the promotion of daily art practice, as art making can contribute to social cohesion, improve one's quality of life, inhibit alienation and keep the imagination alive. Through an understanding of the possibilities for everyday creativity in an extra-terrestrial setting, the participatory activities that engender it and the structures that may enhance or constrain it, this paper explores how art practices might activate and sustain individuals, organisational cultures and communities in frontier societies off-Earth. First-hand contact with a small art based organisation is examined through an alliance formed within my visual art practice: in the past five years I've collaborated with Cameron Noble, an artist from Arts Project Australia, a small community-driven organisation that manages an open studio and gallery for artists with disabilities. Analysis confirms that embedded within the philosophies surrounding the artist studio at Arts Project, are the strategies and structures for good governance of future societies in space, particularly in relation to notions of capacity, development, leadership, diversity, inclusion, ethics and risk. This paper suggests a few simple, best practice guidelines as useful for organisations and communities and applicable to the governance of artists and non-artists working creatively in extra-terrestrial micro-societies. Further, beyond frameworks of need and deficit, it proposes that art making should be mandated and funded as a necessary part of existence for every person living in anomalous settings in frontier settlements far from Earth, set apart from leisure and entertainment and as a part of daily extra-terrestrial life.

Keywords Art · Collaboration · Community · Ethics · Governance · Risk

A. Beattie (✉)
R.M.I.T. University, Swanston St., Melbourne, VIC 3000, Australia
e-mail: annalea.beattie@rmit.edu.au

C.S. Cockell (ed.), *Human Governance Beyond Earth*,
Space and Society, DOI 10.1007/978-3-319-18063-2_14

In the cafe at NASA Ames I spoke with astronomer Duncan Steel about the need for mandated art practices in new extra-terrestrial populations. Given that future settlement, for example on Mars, will involve monotonous, toxic, controlled surroundings, continued surveillance, prolonged isolation and a lack of autonomy, it seemed apparent to me that art making could do much to invigorate and sustain small communities in space. In our conversation, I waved a flag and wondered why art is not considered fundamental to future space exploration? At this point Duncan suggested that individuals probably would make art if they wanted. Of course he was right. Any kind of imaginative practice is a choice. There is no need to formalise anything about art unless associated with personal learning within prisons or similarly, within educational institutions. Mandated art making sounds a bit repulsive, disgusting, not to mention counterproductive to concepts of liberty. But although on Earth it's an accepted human right to participate in art, (United Nations 2014, p. 27) will this automatically be the same elsewhere? I'm always wondering what kind of institutionalised communities will be promoted in planetary exploration and whose beliefs and propositions will have priority? It might be that in the space race in societies far from Earth, art practices could run last, if they are in the race at all (Fig. 14.1).

In any situation governance has its own direction, according to its circumstances. On our planet we understand the structural mechanics of our societies but we don't really know how young societies far away from Earth will proceed. If psychological and social factors can critically determine the success or failure of future space missions, space organisations must identify areas of vulnerability prior to settlement and to minimise future problems, find ways of addressing them. When I think of governance, (it sounds French), illogically but vividly in my mind, I see the grey, black and white of Picasso's *Guernica* (1937), then straight away to horror, to the giants of Goya eating the heads of children. Governance can be related to tyranny and evil, even when intentions are good, enabling leadership and direction, managing people, mitigating risks and compliance with laws. Here on Earth these political goods are based on social beliefs and begin with the supply of security, particularly human security. The other end of the political spectrum is freedom of speech and freedom of expression. These conditions will probably be the much the same in the first settlements in space but in new spacefaring communities, one of the biggest problems associated with governance at a local level might be how to consider who is represented; how to legitimise the roles of different groups, what community means and what objectives underpin its definitions.

What role will artists have in future off-Earth communities? Can our conversations about governance take into account how art-rich cultural practices can contribute to integrated societies, sustainable communities, judicious organizational structures and to the well-being of all humans on distant planets? Apart from space art, how is art perceived within space science communities? Perhaps art is still perceived by its audience as a privileged expression of subjectivity based on creative will, often associated with communities and institutions obsessed with knowledge, power and scale. Or maybe art is seen to be something that is solely self referential and aesthetically introspective. For instance, Robert Nelson explains the

Fig. 14.1 All our discussions about governance begin and end with actions between people. Photographed at our exhibition opening at Trocadero gallery, Footscray, Melbourne, 2013, from the *left*, disability carer Dan Silk and artists Annalea Beattie, Cameron Noble and Tim Noble

twentieth century modernist inclination, 'During modernism, the art work was autonomous, free of contingencies of time and space. No other stakeholders count, from the social and political or economic to family and tradition. The purity of the artist's expressive manipulation of the medium seemed a pre-requisite; and the work was conceptualised unfettered by references. Hence abstraction evolved as the highest sanctity, the apotheosis of process or the self-sufficiency of means' (Nelson 2009, p. 119). And then there are the discolouring power structures of the art market. Differently, and significantly for our discussion about the governance of micro-societies in space, art practices in our frame of reference attempt to situate art firmly in the social world, inventing ways to interpret truth and reality that are premised on ethics, constitutive of common ground. As artists search for inclusive, collective languages that transform perceptions and prompt us to imagine different possibilities for life, meaning is a shared concept and being in the world is shared existence. Concepts of community play a critical and decisive role in this process.

To envisage a future that includes a dialogue about art making and organizational accretion in future inter-planetary settlement, we can only start from where we are, recapitulating human conditions here on Earth. In this case, to broaden the collective process, governance in communities is framed in relation to current models of art intervention; those that address cultural democracy, development and empowerment and those that have the potential to grow long term, collective capacity. Although this conversation advocates for the inclusion of artists in new communities in space, its focus is not whether space agencies need to broaden their agenda or whether governance and emancipation are associated through the imperatives of art. To reckon with how art making might influence governance, the best bet is for me to move close to art practices, to remember my local experiences as an artist associated with a small art-based organisation and then to think carefully about how these ideas translate to making art on the moon or in a small community on Mars, keeping in mind that I am thinking about an area where the organising body may or may not be a constituted legal art based body with an organised board of governance.

Insight into the links between art making and governance come from my own visual arts practice via a partnership formed with 31-year-old artist Cameron Noble. Cameron has spent many years making art in the studio program of Arts Project Australia, a studio and gallery space for artists with intellectual and physical disabilities. In the early seventies APA (Arts Projects Australia) originated as an advocacy organisation to promote the art of disabled people to wider audiences. Within a decade the organisation established a full time studio program for individuals to attend daily. Today APA is staffed by professional contemporary artists, is governed by a organisational management with a volunteer board and is funded mostly through state and federal bodies as well as philanthropic trusts and some additional self-funding. In spite of personal differences and difficulties related to physical and mental health, over a hundred people frequent the artist studio every week, with most attending five days. Unlike space travellers to this point, who choose to go into space and are mostly male, white and physically fit, all sorts of people make up this community of practice. There is a wide range in level of disability, age, and background and nearly all participants begin as non-artists. As it does in any situation, art comes before ability at Arts Project, I'm thinking far away outposts, deep space station. Independent evaluation affirms that, 'Since its inception, Arts Project has sought to provide opportunities and prioritise the process of engaging with art over the consideration of disability. Arts Project sits within a plethora of legislative frameworks around disability but it operates mainly within the world of art.' Netbalance Foundation (2013, p. 4). Much can be learnt from the governance principles of such an exemplary organisation and its community-based artist studio.

As in new extra-terrestrial cultures, dependence, supervision, restriction and a lack of mobility are part of everyday life for artists at Arts Project. By nine-thirty upstairs, in the sunny studio, the atmosphere is buzzing as people arrive. Artists, family members, carers and arts workers move amongst easels and wheelchairs to set up art materials for the working day. Recognising the right for everyone in the studio to enjoy the independence art making provides, relationships in the community are collegial, convivial, carefully civil, as all artists must work side by side, every day, sometimes for years. Drawing and painting are the cornerstones of studio practice though there are plenty of opportunities for other methods of art making including printing, photography, digital animation, sculpture and ceramics. Arts Project bases its charter on a '...social model of disability which focuses on rights not needs, and differences rather than problems' (Arts Victoria 2008, p. 5). This means APA is not an educational facility nor does the studio program focus upon art therapy. Artists specify their practice, there is little directed or instructional activity and apart from the restraints of material and the constraints of disability, nothing is pre-determined. Practices are administered by regular studio review while individual plans set priorities, actions, address obstacles and inform goals within the studio. Careers are overseen and mentored by skilled, qualified, professional art workers, not disability carers. These in turn explicitly inform the role of management. APA leadership responds by supporting studio artists with governance policies that strengthen their position, sustaining them to identify their own needs as artists and maintain art making. Even the physical situation of the studio itself is place-managed rather than program-managed. Underpinning all governance policies and practices at Arts Project are the broad concepts of human rights; respect, dignity, equality, integrity and the basic entitlements of citizenship; equity, inclusion and participation. These well-recognised democratic behaviours are highly regarded within the studio as well as by the organisation, as disabled artists often realise what it means to be vulnerable to exclusion and to abuse.

Cameron arrives a little late with his disabled siblings, Tim and Jodie. He looks subdued. His paramedic carer Dan explains how Cam is still exhausted by a double seizure the day before. Cameron has a heart problem and his epilepsy medication is strong, he does look tired but he is here today to paint with me. Many artists in the studio struggle to continue art making with the physical obstacles of their disability and some, like Cameron, need full time care at home as well as in the artist studio. Dan, as Cameron's disability carer is part of our every negotiation.

Arts Project Australia builds strong identities for its artists through the opportunity to progress and exhibit their work but like big brother in space, interactions with the world outside the studio are sometimes restricted to those APA can provide. Collaborative practices with artists from the wider arts community are sometimes part of this provision. It's time to work and Cameron and I pin up our canvas, choose brushes and mix colour. Johnny Cash is playing in the background and we begin to paint. It's quiet now in the studio, forty artists are focused on their work, just a bit of singing along to Burning Ring of Fire. The task for us today is to finish painting a life-size image of my old VW Beetle with Cameron driving fast behind the wheel, me sitting next to him, looking a bit dour. In the discussions we've had about the future, Cam occasionally mentions leaving Arts Project but because of his health, there are no real options available. Conversations about mobility and its lack have motivated the direction of our subject matter and the context of our collaboration.

Though he will never drive, in this image Cameron is in control, intent on the road ahead, spinning to the left, leaning out the window and adjusting the mirror casually as we speed along. This wishful portrait of us driving away is the culmination of a series of assumptions and methods of enquiry that reflect particular stages of the work process. Through observation and trial and error, we've experimented with different materials, setting up tools for thinking and changing thought, shifting course to find mutual ground. We like not knowing what will happen and part of the potential in our art making is that we can change tactics and engage differently because of our failures. There's some rhythm in it. For me, the resulting painting on one hand feels purposeful, a bit driven itself. On the other hand, the final product is residual, almost negligible. Deliberate art practices are always transitional in intent and ongoing in outcome because the goal is found within the sustained activity itself. When the art making works well for us, the process and the concept are indivisible. These actions are always true to their emerging material conditions and mutating context, not more of the same.

Thinking of what might happen if art practices were an intentional part of working life for young frontier communities in space, I am reminded of my limited experiences of a simulated space environment at the Mars Desert Research Station in Utah. In that space analogue environment, when the weight of daily activity was geared towards safety and survival, mission crews had little time for anything else other than negotiated duties. By the end of the day everyone was physically and mentally fatigued. When days are so overwhelmingly busy, the continued effort and time necessary for works of art can be hard to summon up. When examining the role of deliberate practice, behavioural psychologists recognise that in any context it is crucial to monitor the necessary degree of mental, physical and emotional commitment, frequency and duration and the level of reception and support, claiming, 'Disregard for the effort constraint on deliberate practice can lead to injury and even failure. In the short term, optimal deliberate practice maintains equilibrium between effort and recovery. In the long term, it negotiates effort constraint by slow, regular increases in amounts of practice that allow for adaptation to increased demands' (Anders et al. 1994, p. 371). Actually, intuiting methods to deliberately create art calls for much more than building up one's stamina by balancing time, effort and recovery. Art making requires you explore an unfamiliar part of yourself, to bring to light something of the courage of your convictions. Risks are taken in this low-stake environment but there are no Flight Rules, no functional formula, no standard operating procedures, no behavioural psychologists and the world created by art making is always incomplete. The challenge is to find your way. Recognising what you find can sometimes be confronting.

When I spoke to Arts Project director Sue Roff about good practices for art making in the artist studio and the impact of these on the actualities of governance, she stated that, 'Ethics is a prime consideration for all stakeholders in every circumstance' (2014, Personal Comm.) In small reliant communities, the limit of obligation in any context is an ethical question. For instance, how to maintain integrity whilst avoiding lowest common denominator decisions is a key issue for collaborative and participatory art practices. When Cameron and I are painting, I notice ethical behaviour falls in directly behind art making; as a caretaker supporting observation, quietly scaffolding methods of possible investigation, assisting the art work to proceed steadily, with rigour though not necessarily always in an even or horizontal manner. Ethics arises to regulate and preserve the realm of truth through subjective commitment to specific moments, events and encounters within singular situations, all of which philosopher Alain Badiou names as starting points for human action. Badiou positions truth procedures such as ethics through a process of fidelity within an objective situation, inducing subjectivity through a response to an event: '...that which this fidelity produces in the situation' (Badiou 2001, p. 43). In art making, this crisis of meaning, of being true, holding true to a principle, person or event mediates our intentions upon a state of change. On any ordinary day our collaboration can involve disappointing false starts, stubborn retracing of steps, desperate searching for the faintness of methods, disorientated wrong turns and contrary struggling to maintain our equilibrium as well as our momentum when the work founders. At the same time, when rare singular moments of connection occur, these illuminating encounters shine ecstatically bright, reconciling matter, thought and existence in a such a way that surpasses any visceral satisfaction involved in making art. Experiential junctures like these seem far from the principles of scientific research and partly, that's their strength of purpose in this discussion.

Cameron and I are at odds over painting method. I am convinced that to be able to see the gesture of the hand in the brushwork is a good idea. We need to give the work some texture, I argue, so that when the painting is reproduced as a paste-up and exhibited on the wall of a busy corner, its painterly qualities are emphasised, the paint is acknowledged in the printed image, showing process, referring away from the support, showing the root plasticity and so on. Cameron disagrees. Unlike me, he isn't influenced by modernist art training and he doesn't care less about the paint or process. Fair enough, he intently and deftly brushes out every thickness of paint, every rise and fall of colour, graphically blending in not only his marks, but mine as well. Full of muscle memory, Cameron moves around our image like a dancer, quickly painting over my work. I feel crotchety but what can I do? I don't agree with him but I defer as Cameron is a determined negotiator. There are plenty of hiccups and minor skirmishes like this one but as Alain Badiou says, 'Continue! This is the maxim' (Badiou 2001, p. 47). While Cameron and I try to figure out what we are doing together, more likely we proceed erratically, bumping along one step at a time. As things speed up, there is insouciance. We have tangible confidence in our partnership, we are artists, we can speak in different registers, we have hunches and we are shrewd. It doesn't matter at all if we think differently, our outcomes agree. The negotiated aim for this project has remained firm: our collaboration focuses on issues of mobility and an exchange of portraiture. Our desire is to extend the boundaries of our art practice beyond the limits of the studio by engaging with public art and the demands that domain presents.

I keep thinking about one particular well-known collaborative land art project when I think about art, ethics, restricted communities and moments of shared truth. Francis Alys's *When Faith Moves Mountains* (2002) is the story of five hundred Peruvians from a small, dispossessed pueblo community in Latin America, who were paid to move an enormous sand dune ten centimetres forward, one shovel of sand at a time. Alys mobilised people living in an immigrant town to make a long human chain on the dunes and move forward, digging and displacing the ground beneath them. In a place where, on a daily basis, dislocated people endure continual political crisis, what kind of proposition is present in such an incongruous work with its emphasis on material deprivation, displacement and manual labour? Ferguson and Fisher (2011) suggests that the ethical responsibility of *When Faith Moves Mountains* is found within art that has a story telling function, one that sees the encounter as collective action, summoning feelings and the imagination that are to do with poiesis and subsequently liberty. Fisher says, 'In repressive regimes, the cruellest act of all is perhaps not to deprive a people of its body but of the ground and will to imagine new possibilities of life' (Ferguson and Fisher 2011, p. 119). Alys chooses to work with myth and narrative in a way that leaves behind the story of a common threshold moment and nothing else. This potential to dream and imagine a different future is the premise for the social allegories he creates and the key to art making in extra-terrestrial societies.

It's mid-morning on a Tuesday and I'm at the Arts Project studio. I'm waiting for Cameron while he wastes a bit of time chatting to his mates. There's a bit of racket, he seems to be stirring up trouble. While I wait, I lean on the sink cleaning his brushes, which is not okay. As I look about, most people are absorbed in their own art making. I'm still thinking how the goals in the studio inform the directions taken by APA management and how that might translate to an extra-terrestrial community. On the other hand Cameron, like most 31-year-old men, is oblivious to whether the idea of governance is an artefact of community or whether community is a technology of governance. Yet like all artists at APA and artists elsewhere, he recognises that he is part of a group of people for whom the subject, method, context and outcome of art making is the art practice. His identity as an artist and the significance of art for him and for me, comes through making art that is ideally embedded in everyday discourses, refined and affirmed within the socialness of communities. It's not an ideal world but everyone at APA knows that there is a self-sustaining ethos present. Meaning created in the artist studio is calibrated, validated and manifest within a group of like-minded individuals. When we think about art making in small confined settings like this one, Biggs and Bucher suggest, 'There is a relationship between a community's value system, the activities they deem relevant and the actions that comprise those activities. When this relationship is coherent, the community is satisfied because their values have conditioned the actions and are therefore reflected in them' (2011, p. 85). In these political terms, values and beliefs align with the indeterminate activities and experiences of art practice through flexible networks within a trusting community. Simply and bureaucratically, this instrumental tying together of associations, in a place where art is valued and respected, identifies transparent human-centred outcomes which are then expanded to create norms amongst all stakeholders. This leads to better organisational structures depending on whether these norms are enhanced or hindered by governance.

This said, like any group of individuals, the APA studio artists may share reasoning around structuring their experiences through art making but their community does not necessarily hold the same moral beliefs or ideals. How does a community with art in common encompass difference, whilst drawing people together? Similar to networked science communities, artists participate in their community of practice with the aim of bringing about new disciplinary discourses through critical reception. Unlike science, consensus is not a priority. Judgement is found within art through material reflection and dissensus. While art exercises a conception of self, '...its ethical dimension is the purposive nature of its indeterminacy, which invites a kind of communication that leads to shared terms for norms and values. 'Leads to' is the operative phrase here, as there is no actual necessity that we arrive at a consensus. Consensus as an ideal can structure the communicative process without ever actually occurring. After all, a firm, stable long-lasting actual consensus would be inimical to cultural renewal' (McMahon 2014, p. 74).

If we consider this idea, it's important to think about what kind of cultures will be perpetuated by carefully screened space crews in new frontier communities in space. Will there be a place for contradictions and fissures in belief within these bounded groups? In subsequent generations in space, will the ethics of small uniform space societies be called into question by such concepts as uncertainty or ambiguity, not to mention irreverence, failure, whimsy or desire? Creativity is charged by these concepts within communities who practise art.

To achieve good governance through inclusive communities and positive change requires more than internal reciprocity. A balance between a democratic framework and the development of a community who holds art making in esteem needs to be supported by policies at a local and federal level. To open up these relationships to scrutiny and to appropriate modification, evaluation is an important factor in any strategic approach. What could be used to evaluate the impact of art making on tiny, closeted communities living in isolation, for example, on the moon?

On Earth, Quality of Life is recognised by the World Health Organisation (2002) as a 'non-stigmatising concept that reflects whether an individual is suited to her or his environment, without applying judgement on capacity.' Recognising context as a conceptual framework for the evaluation of complex outcomes, Q.O.L. encompasses personal development, material and emotional wellbeing, social interactions and self-determination. Likewise the Australia Council (2011) as peak organisational body for artists in Australia, affirms that enhancing quality of life is a major goal for art-based communities. In line with these organisations, and as a way of understanding individual and collective outcomes at broader levels of government, the Productivity Commission for Disability Care and Support (2012) acknowledges Quality of Life as a person-focused measure that is increasingly being used to coordinate, deliver and evaluate services. This transparent standard of engagement gives priority to the needs and interests of the people it serves and motivates organisations to ensure that the means and processes are available to reinforce common goals. As it seeks a rich narrative based on personal and communal experiences of change, Quality of Life is a useful way of thinking about art-related outcomes in small community-driven systems of governance.

At Arts Project Australia, the studio creates social value for all stakeholders and as a small-scale organisation, Arts Project has demonstrated that open forms of art making directly address quality of life, providing positive social value. 'At APA artists are the major beneficiaries but the studio has created social value for all stakeholders' (Netbalance Foundation 2013, p. 23). Like scientists, artists recognise we are at our social best in communities we value and who value us, as well as our work. This is not a new model. As Kant suggests, this is 'sensus communis—aesthetic, reflective judgement that is dynamically and minutely inter-subjective: it tacitly assumes a communal orientation and is relative to the values of the community in which it is formed' (Kant 1790, pp. 291–302).

When we think about leaving Earth to live and die in a place that lacks the most fundamental physical freedoms, the most basic being to breathe fresh air, then quality of life is something we immediately understand as contextual. Living inside in a cloistered community under constant surveillance in barren, hostile surroundings can easily diminish one's quality of life. The ennui and lassitude that might increase over time is a major problem for human factors in extra-terrestrial settlement. Tonight Mars in the sky is 34.6 million miles away. I'm still worried about the kind of strange, raw micro-societies that in the future will coalesce in the crannies of its thin dusty atmosphere. Over generations, will ethical decision-making eventually rely upon the unfettered power of a few idiosyncratic individuals? What's missing from society in such a place will be just as important to people as what is there. Will our future in space be a bizarre, dystopian world inhabited solely by scientists, technologists and technocrats? Such crypto-fascist microcosms are not viable and doomed to fail.

Through shared goals cognized through community, art offers governance different tools for thinking and more importantly, different ways for being. Isolated off-Earth communities do need art, music, writing and other forms of creative behaviour to nourish the imagination and curb dark sociopathic wolves at the door. It's not enough to construct the means to leave our biosphere and travel into space. To conceive of new, sustainable ways to live in other worlds, space travellers must have the power to lose themselves and be lost in the universe. On Mars, where sunlight is deadly, the air is poisonous, where small groups of humans will be sealed together in stale containers, when every moment of time is scrutinized and accounted for, individuals might make art if they feel like it. However, my belief is that if a conception of an arts community is not prioritised before the first settlement in space, if artists as professionals are not included and if art practices are not mandated and funded as part of daily life, then leisure and entertainment will take their place (Fig. 14.2).

Some suggestions for governance in an extra-terrestrial community:

1. To promote ethical and responsible decision-making and to lay solid foundations for management and oversight in small off-Earth communities, artists should be included from the initial stages in the formal processes of political engagement.
2. As on Earth, in extra-terrestrial environments, every person should have the right to sustain art practice and be free to choose their own subject matter, form and content within the constraints of materials, ethics, health and safety, funding, time and site specificity.

Fig. 14.2 Artist Cameron Noble outside Art Projects Australia, painting a fast car on a slow car. June 2013

3. To protect and promote freedom of expression, the conditions of creativity should be acknowledged as being able to act without coercion and censorship, with spontaneity and foresight from a position of certain psychological safety.
4. To manage risk and improve quality of life in an extra-terrestrial setting, art practice needs to be acknowledged as positively impacting upon personal development, material and emotional wellbeing and social interactions.
5. To foster extra-terrestrial organisations that promote diversity and inclusion, art practices should be recognised as accessible to every person regardless of ability and acknowledged as transitional in intent and ongoing in outcome.
6. To strengthen cultural capacity in sustainable communities off-Earth, organisations should responsively enable artists and non-artists alike with the means and processes to identify their own needs in order to maintain art making.
7. To respect the legitimate interests of all stakeholders in art making in new extra-terrestrial societies, deliberate art practices should be mandated, resourced and funded, set apart from leisure and entertainment and as part of the daily experiences and activities that constitute every person's communal working life in an anomalous setting.

References

Anders, E., Krampe, R., & Tesch-Romer, C. (1994). The role of deliberate practice in the acquisition of expert performance. *Psychological Review, 3*, 363–406.

Arts Victoria. (2008). *Picture this: Increasing the cultural participation of people with a disability in Victoria*. Retrieved 21 April http://www.arts.vic.gov.au/research.

Australian Council for the Arts. (2011). *Essential governance practices for arts organisations*. Retrieved March 10, 2014, from http://www.australiacouncil.gov.au/__data/assets/pdf_file/0009/108378/Australia_Council_-_Essential_Governance_Practices_for_Arts_organisations.pdf.

Badiou, A. (2001). *Ethics: An essay on the understanding of evil*. (P. Hallward, Trans.). London: Verso Books.

Biggs, M., & Buchler, D. (2011). Communities, values, conventions and actions. In *The Routledge companian to research in the arts*. London: Routledge Press.

Ferguson, R., & Fisher, J. (2011). *Francis Alys*. New York: Phaidon Press.

Kant, I. (1790). *Critique of judgement* (W. S. Pluhar. Trans.). Indianapolis: Hackett Press.

McMahon, J. (2014). *Art and ethics in a material world*. New York: Routledge Press.

Nelson, R. (2009). *The jealousy of ideas*. Melbourne: Ellikon Press.

Netbalance Foundation. (2013). *The social impact of arts project Australia*. Retrieved March 20, 2014, from http://www.artsproject.org.au/content/social-impact-evaluation.

Productivity Commission. (2012). *Disability care and support, report 54*. Retrieved March 18, http://www.pc.gov.au.projects/inquiry/disability-support/report.

United Nations. (2014). *Universal declaration of human rights*. Retrieved April 21, 2014, http://www.un.org/en/documents/udhr/index.shtml#a27.

World Health Organisation. (2002). *Toward a common language for functioning, disability and health*. Retrieved April 29, 2014, http://www.ncbi.nlm.nih.gov/pubmed/8560308.

Chapter 15
Are We There yet? Pursuing a Utopian Ideal in Education in Space

Janet de Vigne

Abstract Many people are currently considering education in the 21st century as something quite different from what we have been used to. Some, like Jenkins (2002), consider that it is now outside the classroom and should not be 'controlled' in traditional ways. Others, such as Mitra et al. (2013), consider education as an 'emergent' phenomenon, something that will, with a little less help than usual, become a very effective self-organising structure. With the advent of two new intelligences as advocated by Gardner (1999-existential intelligence and naturalist intelligence), how could education in the closed system of a spaceship or extra-terrestrial settlement take into account new thinking and the connectedness of humanity in its acquisition and practice? How might teachers function in this environment and how much might depend on the computer? How might AI emerge as a powerful factor in the development of human beings carrying the future of the species into hostile environments? These questions and more are considered from the perspective of education—a human right? And a child's freedom to make choices concerning her future, her education and a 'career', in terms different to those currently understood.

Keywords Education in space · Educational liberty · Educational technology · Utopia

Everyone, it seems, has an interest in education in the 21st century; at the very least, they can tell you that everything seems to be wrong with it and that the system is failing our children. In examining how the education system might function in the confines of a space ship on its way to a habitable planet, this paper will present contemporary ideas shaping learning and teaching, while exploring the very specific needs presented by such a case scenario. It is hoped that a better understanding of

J. de Vigne (✉)
Education and Social Justice, University of Lancaster, Lancaster, UK
e-mail: janet_devigne@yahoo.com

C.S. Cockell (ed.), *Human Governance Beyond Earth*,
Space and Society, DOI 10.1007/978-3-319-18063-2_15

today's classroom will emerge, and that a common interest in improvement and development will ensue. To quote Dr Mae Jemison, 'We can go to the village because we've been to the stars'.

15.1 The Context

When we consider the fleeting nature of human affairs, the restless and uneasy spirit of our times, when every generation overturns the work of its predecessor, can we conceive a more senseless plan than to educate a child as if he would never leave his room? (or… planet?) (Rousseau 1762; Gardner 2000).

Assuming that humanity develops the capacity to leave the earth in search of another habitable planet, and that the speed of travel to such a planet will not exceed the laws of physics currently understood (in other words not faster than light), survival of our species will depend on our ability to produce and raise children in space who will need two distinct groups of skills. The first group will involve the operation of very sophisticated machinery in order to survive within the spacecraft and reproduce and manage perhaps two more generations on board. The last (space born) children will need a second group of skills to equip them for landfall and survival in a set of circumstances almost impossible to foresee. It may be possible to design a system that will develop with them, (i.e. an education system not entirely dependent on human interaction) therefore one of the most important issues here will be 'critical mass', not just because of the numbers of people necessary to continue the species, but because of the current understanding of emergence and the 'self-organising structural' elements of human consciousness and understanding.

In the words of Cockell (2013), children in space will be brought up in cages. The cage of course has a dual purpose—it keeps the children inside a safe environment, and it prevents them accidentally or deliberately entering or exposing the community to a potentially lethal one. The issue of nature or nurture will become contentious; teenage rebellion becomes a serious threat to life, the curious toddler becomes a hazard not just to herself but the entire community. Skills that will need to be taught will have to encompass reflection, understanding and management of the self, responsibility and accountability to the community and, very importantly, conflict resolution. This is of course no guarantee at all of a non-autocratic 'one big happy family', but the settlement members (or crew) will have no option but to take the calculated risks of growing 'community' in a container where no-one can hide (or not for long) or leave (during their lifetime). And we have not yet mentioned subject knowledge.

Children in space will need to learn fast. They will need functioning maths, astrophysics, biology and who knows how much science in order to make their environment work for them, exercise control and repair of highly specialised equipment. An in depth understanding of what they start with will enable them to develop and innovate while in space, thus improving the environment for the community. Creativity and freedom to think, problem solving and practical strategies for invention can be taught—not using the traditional jug of water and empty glasses teaching metaphor, but something much more sophisticated—the teacher as enabler, gardener, guide? However necessary these subjects might be, where might they leave the learners in terms of artistic development? Linguistic development? Human development? What does current neuroscience have to say about these subjects, largely ignored by those panicking about STEM? And could a programme be developed that would create a utopian educational environment, prepared using what we know now, and designed to expand and develop to encompass what we do not yet know? Where are the child's choices in all of this—and how can we guarantee that she will be able to develop along these lines?

Let us briefly examine the reasons why we educate and have educated our young people today. The game has changed.

From the 1100 s and the beginnings of the great European universities emerging from the mists of the post Roman era, education developed within a very strictly controlled system. Galileo, Copernicus and many more were persecuted for disagreeing with the religious state in matters as basic as fundamentals of physics and maths. This seems remarkable to us, so far away from the beliefs of that early church. And yet what people were allowed to access at that time was rationed—an education fitted a student to become a priest and debate weighty theological matters (usefully such as how many angels could stand on the head of a pin), to take the office that was his by right or by might, or to trade internationally. Knowledge created a division between the elite and the non-educated, perpetuating myths of 'divine right' and the fatalistic acceptance of the underclasses of their supposed inferiority by divine design. (Is this ringing any bells? Social mobility was non-existent.) Myths of the torments of hell, pornographic in their visceral detail of punishment for sins such as adultery, were legion. Moving through the Renaissance and in England in particular, the state wrested control of education from the church, and access to Oxford and Cambridge was allowed only to members of the Church of England, thus contributing to the movement of scholarship between Europe and Scotland, reinforcing Scotland's separate identity as a nation state. As the Reformation swept through Britain, it became necessary for the 'moral good' to allow men and boys to read the Bible—in their own language—to maintain religious civic control. Gaining some rights, partly though being able to read, people were prepared to die to defend the same. Moving on again by the time of the Industrial Revolution, education begins to provide a mechanism of economic control—everyone should be able to access the instructions for the operation of the machines, not for their personal safety, but in order to maximise profit for the factory owner. We then move forward again through economic stages one, two and three where education becomes more necessary in terms of production and to some degree protection of the workforce (for sound economic reasons rather than altruism, such as income tax), litigation procedures as the workforce becomes 'aware', until we reach today's ideologies of economic power and the knowledge economy. Fourth sector—and no-one knows how it will pan out, because we in the West, the traditional leaders, have not been here before. So—what is happening? Knowledge is increasingly within the grasp of anyone who can access the Internet. At least, quite a lot of knowledge is. This is of course a ridiculous and somewhat cynical oversimplification, but it is not without some justification.

Those striving for Utopia have, until now, conceived a reductionist context, reductionist because it is so rigidly controlled. Some would remove children from their parents, some would structure a town in a particular way to facilitate the smooth running of the new society. Some argue that Utopia cannot exist without its dystopia, again using a binary reductionism that in the 21st century does not sit comfortably with developing concepts of emergence or quantum, chaos and complexity theory.

15.2 Education and Knowledge

Life is not breath, but action, the use of our senses, our mind, our faculties; every part of ourselves which makes us conscious of our being. Life consists less in length of days than in the keen sense of living (Rousseau 1762, op cit).

Education might indeed be intended to equip a child for life—in this context, certain uncontrollable and not agreed-upon circumstances may well render a child less capable of exercising certain of her human rights. A new paradigm of when and how certain rights may be attained and which are inviolate from before birth will be necessary—today, these issues are the subject of much international discussion. Before children were considered as people, Rousseau argued passionately against certain practices including swaddling, common in 18th century as a way of physically containing young children. Taken to extremes, it was abused as a way of controlling otherwise bothersome behaviour. For Rousseau, 'bothersome' is the very means of interacting with the world that allows a child to learn. In some parts of this immensely important treatise, Emile receives an education that does express his liberty. But in others, he is just as much put upon as any other child, as what he 'needs to know' is decided for him. Considering the education of the Wart (the child who will become Arthur, the once and future king), White goes much deeper into an interaction with the world. Through the gifts of the magician Merlin, Wart is able to become a fish or a bird (White 1938). These transformations place him in real danger, (in these experiences he is unaware of the presence of his teacher as a protective force) but the learning curve is huge. (Understanding the world from an animal's perspective is a concept being explored currently in education.) Could White have been predicting a virtual world where such profoundly subjective experiences might be attainable through non-magical means?

I am equating knowledge acquisition, rightly or wrongly with education, as most of the types of knowledge outside survival skills associated with terrain or culture are acquired through practices of literacy such as reading, or using a computer. There is no reason why one should not be able to educate one's self—history is littered with highly motivated people who have done just this—but it is unusual. In fact, according to Jenkins, education is now out of the control of the state, the church and the classroom and is within the grasp of the people. One consequence of this is democratisation, another lightning fast innovation as ideas travel between countries. Is this a happy place for policy makers? How might such a paradigm affect an education programme designed for space?

Policy makers have reasons to rejoice and reasons to fear current developments. Monitoring people's behaviour and thinking has never been so easy. As fast as Facebook-style social media sites emerge, so does the technology to 'observe' interaction thereon, enabling both trade and political manoeuvring. The role of such sites in revolution and the communication of disaffection has been well noted. If Jenkins is right and the reign of the new literati of the Internet is just beginning, then ancient institutions holding power had better look to their laurels. I suggest that it is in the nature of these institutions to survive, as they have done for centuries, and that their capacity to adapt may surprise the revolutionaries. The universities for example, in producing Coursera and other Mooc delivery platforms are already investing billions in systems to educate the world. You pay for the certificate at the end of the course, not the education, which will contribute in equal parts to a new equality and inequality in society—to be certified or not might well be the first issue on the application form. The capacity of these courses is huge; hundreds of thousands of students worldwide, learning, talking and giving feedback on the teaching, changing it to meet their needs. A truly flexible and complex system arises therefore, where liberty is enacted constantly.

So the issue here pertaining to space may be one of 'connectedness'. Humanity connected around the globe, togetherness flattening society and producing a best case scenario Utopian democracy in Jenkins' brave new world (op cit, Jenkins 2002). (It might not of course—there are those who fear deception and tyranny arising through the same mechanisms.) The advantages of this collective intelligence are clear—education, now outwith traditional contexts (the classroom), can truly be owned and therefore driven by those who are learning, and not by those who have a vested interest in what they learn. It might become a truly dynamic system where the learner and teacher ultimately become inseparable. Therefore, according to Jenkins, some of the best things educators should give students today might be the social skills and cultural competence to function in a networked society. In space, the connectedness will not be there and there may as a result be a threat to individual liberty as a consequence—if liberty is a construct held, albeit loosely, by the many. The possible beneficial outcomes of emergence, the rise of the self-organising structure, the glorious messiness of the mass of humanity will have to be vicariously experienced via the computer data banks. It would at least be possible to include vast amounts of information in increasingly small spaces. But are there any instances of self-organising structures in education today that might help us understand, support and build a dynamic system in space?

The research of Mitra (2012) bears investigation. An educational technologist by training, Professor Mitra believes he has stumbled on a technique of allowing schoolchildren to engage in a teacherless system that illustrates a self-organising structure in education. This is important, because it means the system is greater than the sum of its parts. Children are placed in SOLEs—self organising learning environments and given problems or questions to solve or answer. Left to their own devices, with adults there just to admire them and encourage them, children are capable of learning (a) more than we think they can (within supposed developmental constraints) and (b) retaining substantially more and more advanced information. This has surprised many teachers—but has been foregrounded in education by the work of Vygotsky and Bruner in 'scaffolding'. Arguably, Mitra's work relies on 'situational memory'—the child recalls instantly who she was with, when and where—and all this contributes to the retention of the knowledge gained aand arguably the liberty of the mind to function at its best. MItra's TED talks are exciting and inspiring—in fact he won the 2013 TED prize for his pioneering work in the field and has been given £1 million to continue his research at the University of Newcastle, UK.

So, access to information, while it will not be real time, is possible, but there are certainly stories of children using the Internet only as a source and being led astray by the first findings thrown up by the search engine. In space, accessible information will not be live, changed and changeable—it will become fixed (unless we can create a dynamic system somehow) and one of the most important aspects of Mitra's concept will be missing—the Granny Cloud—the adult enablers who stand behind the children and 'admire' them. Enabling children to teach themselves seems to offer many solutions to the logistical problems of an education in space, even freeing the learning capacity of the students, but it doesn't appear to solve issues of freedom of choice, of artistic development and emotional development (such as conflict resolution). What will happen if you know you are born to be a dancer? If you are gay? If someone imposes their will on you? If you fall in love and it is not requited? So it seems that 'significant other' adults will need to be equipped to negotiate these topics. This is highly skilled work and requires specialist training, but there would still be no guarantee that a 'good' education would result in the survival and prosperity of the species. Might the space travellers need an extra element of control in the classroom, at a level that we would, today, find ethically unacceptable, without going down the road of the trans-humanists?

Professor Patricia Kuhl (2004) is considering brain based interventions that could 'help' a child with difficulties in maths, or maintain the 'plasticity' of the brain to allow it to keep growing rather than fossilise (the Lenneberg hypothesis of 1967 has, it seems, been confirmed by her research, in that the brain's capacity to absorb new languages seems to stop at age 12 as the brain takes the shape it will have in the adult). Kuhl has shown that babies are statisticians—by measuring their reactions to language she has observed how the brain calculates instances of repetition of sounds and stores them in recognition centres. Interestingly, the babies in her research do not respond to artificial stimuli such as a television or audio system —they will only learn from a human. This raises other issues in space—how much stimulation could be provided? What needs to be human and what could be machine? When and how could a child experience the smell of newly mown grass? The sweetness of a horse's breath? In the same way that a no-gravity context lessens bone mass in the legs because they aren't being used, might a lack of olfactory stimulation cause the travellers to lose their sense of smell over three or four generations? This would have huge repercussions at landfall as an essential part of our reptile brain instincts.

We must begin to consider these issues now, as people prepare for humanity to leave planet earth. Education is the single most important element to establish for any human future. Unfortunately, in many contemporary societies, teachers are not valued highly—in fact, they feel they are despised. We are still at the stage where Brendan Iribe's holodeck (the Oculus VR, Young 2014) is considered by some as the Next Big Thing and the saviour of the education system as we know it. Could this provide a 21st century technological Merlin to deepen and subjectify education in areas as yet almost impossible to quantify? Howard-Jones (2014) and Howard-Jones and Murray (2003), at the University of Bristol, would beg to disagree. Researching the effectiveness of educational games and technology in education, he is asking serious questions based in neuroscience about learning, how it happens, and whether such games have a positive or negative influence. What seems to work in terms of engaging learners' motivation, is 'uncertain reward'—to gamble 'double or quits' at the end of the game. This seems to increase dopamine production—the feel good hormone. If you feel good, you are more motivated to engage with your learning. Howard-Jones has showed that the main issues with learning appear to be: enough sleep, looking at the areas of your own brain that light up when you're engaged in the desired activity, and stimulating your own creativity by focussing and defocussing on the task in hand (Howard-Jones, op cit). Certainly the jury here is out on whether the use of certain types of technology can help. One could argue that this might be because the technology in the classroom is very far behind the technology in the home, and that as a consequence children see themselves as walking into a pretty much useless museum of antiquity when they walk into school, questioning whether what happens inside the classroom could possibly be of use to them outside. This means that, in terms of providing a space where learning is immediately engaged, teachers today are at a disadvantage, having to waste time either ignoring or proving relevance before anything can be achieved. Currently, teacher literacy in the area of cross media education is being investigated.

In the closed context of space, though, as much consideration must be given to the teacher as to the taught. Who is going to teach? In what medium—in terms of platforms (technology etc.) but also language—have we assumed that English will be the only language of instruction? Who will the students be? Where will they have come from? Surely not just the West? Plurilinguals are being shown to have an educational advantage in the classroom—but this is not just about brain function. Concepts of empowerment, dissent and dialogue of both student and teacher may be addressed through critical pedagogy; making meaning through language is where hope for the human race lies, not just in but important for negotiation of conflict:

Competencies for intercultural interaction entail the capacity to deal critically and successfully with dissent and even critical cultural awareness towards the Self and the Other through honest and balanced negotiation (Phipps and Guilherme 2004).

Negotiation and resolution of conflict is perhaps one of the greatest liberties we might be able to enact. The will of the many is honestly and fairly debated and a compromise is reached. How important might this be in connecting with other life forms and/or respecting other environments? Theorists who argue for these methods believe in the essential goodness of humanity, in other words that if presented with these ideas, children will choose to be good. Again, we invoke Rousseau—that man in his natural state must become corrupt—he does not begin that way. In a truly Utopian ideal, then, children would actively enjoy, want to engage with and remain uncorrupted by the practices of the society around them, happily complicit, willing and free to engage in their learning.

Most educational theorists agree that motivation is the key element in learning. Types of motivation are legion (this is a huge area of research) but a concept connected to this is how to engage the 'best' kind of motivation—could it be by appealing to a learner's intelligence 'package', as the optimal way in? Howard Gardner (op cit) now believes that there are around thirteen different types of intelligence, and that manifesting seven or eight of these is 'perhaps a working definition of the species'. Gardner refers to leadership, moral excellence and creativity as fostering human potential—qualities certainly desirable in the development of our space children, if a colony is to survive. And thus we return again to the question of control. Can we afford to trust the process of human development? If we try to control this, what might the consequences be? What will need to be learned? And how will it be learned? In the words of Socrates, 'Education is the kindling of a flame, not a vessel to be filled'. Who will teach? What will they teach and how? There are many instances of alien life forms from Star Trek to The Fifth Element being appalled by finding the encyclopaedia entry for 'war' and almost giving up on the human race as a consequence. The ongoing BBC TV project 'Up', which has been following a group of children growing up over a period of 20 years (this predates the Truman Show) generated the following anonymous post a few months ago:

1 'The futures of children are not preordained. They have lives full of potential, luck (or lack of it) and opportunities…'
2 In our case, the opportunities for the first few generations (on the space ship travelling toward the nearest habitable planet) will be very limited; those for the landfall generation absolutely limitless. We are very far from a Utopian ideal. The real question will be, as very possibly in many things, how much can we truly afford to leave to luck?

References

Cockell, C. S. (2013). *Essays on extra-terrestrial liberty*. Shoving Leopard: Edinburgh.

Gardner, H. (2000). *Intelligence reframed—Multiple intelligences for the 21st century*. New York: Basic Books.

Howard-Jones, P. A. (2014). *Neuroscience and education: A review of educational interventions and approaches informed by neuroscience*. London: The Education Endowment Foundation. www.educationenvironmentfoundation.org.uk.

Jenkins, H. (2002). *Interactive audiences? The collective intelligence of media fans*. labweb.education.wisc.edu.
Kuhl, P. (2004). Early language acquisition—Cracking the speech code. *Nature Reviews Neuroscience, 5*, 831–843.
Lenneberg, E. H. (1967). *Biological foundations of language*. New York: Wiley.
Mitra, S. (2012). Children and the Internet—A preliminary study in Uruguay. *International Journal of Humanities and Social Science, 2*(15), 123–129.
Mitra, S. et al. (2013). Self Organised Learning Environments (SOLEs) in an English School: An example of transformative pedagogy? *Online Educational Research Journal*, 1–19.
Howard-Jones P. A., & Murray, S. (2003). Ideational productivity, focus of attention, and context. *Creativity Research Journal, 15*(2–3), 153–166 Taylor and Francis Online.
Phipps, A., & Guilherme, M. (Eds.). (2004). *Critical pedagogy: political approaches to language and intercultural communication*. Clevedon: Multilingual Matters Ltd.
Rousseau, J. J. (1762). Emile via www.projectgutenberg.org.
White, T. H. (1938). *The sword in the stone*. London: Williams Collins Sons & Co. Ltd.
Young, J. R. (2014). *Brendan Iribe and the Oculus VR Chronicle of higher education*.

Index

C.S. Cockell (ed.), *Human Governance Beyond Earth*,
Space and Society, DOI 10.1007/978-3-319-18063-2

MIX
Papier aus verantwortungsvollen Quellen
Paper from responsible sources
FSC® C105338

If you have any concerns about our products,
you can contact us on
ProductSafety@springernature.com

In case Publisher is established outside the EU,
the EU authorized representative is:
Springer Nature Customer Service Center GmbH
Europaplatz 3, 69115 Heidelberg, Germany

Printed by Libri Plureos GmbH
in Hamburg, Germany